HOMEMADE
LIVING

Keeping Chickens

with Ashley English

HOMEM🏠DE
LIVING

Keeping Chickens

with Ashley English

All You Need to Know to Care for a Happy, Healthy Flock

LARK
BOOKS

A Division of Sterling Publishing Co., Inc.
New York / London

Senior Editor: Nicole McConville

Editor: Linda Kopp

Creative Director: Chris Bryant

Layout & Design: Eric Stevens

Illustrators: Orrin Lundgren, Melanie Powell, and Eric Stevens

Photographer: Lynne Harty

Cover Designer: Eric Stevens

This book is FSC Chain-of-Custody certified and has been printed and bound in a responsible manner using recycled materials and agri-based inks.

FSC

Mixed Sources
Product group from well-managed forests, controlled sources and recycled wood or fiber

Cert no. SW-COC-000952
www.fsc.org
© 1996 Forest Stewardship Council

Library of Congress Cataloging-in-Publication Data

English, Ashley, 1976-
  Homemade living : keeping chickens with Ashley English : all you need to know to care for a happy, healthy flock. -- 1st ed.
     p. cm.
  Includes bibliographical references and index.
  ISBN 978-1-60059-490-8 (HC-PLC : alk. paper)
  1. Chickens. 2. Poultry farms. I. Title.
  SF487.E573 2010
  636.5--dc22

                          2009021366

10 9 8 7 6 5 4 3 2 1

First Edition

Published by Lark Books, A Division of
Sterling Publishing Co., Inc.
387 Park Avenue South, New York, NY 10016

Distributed in Canada by Sterling Publishing,
c/o Canadian Manda Group, 165 Dufferin Street
Toronto, Ontario, Canada M6K 3H6

Distributed in the United Kingdom by GMC Distribution Services,
Castle Place, 166 High Street, Lewes, East Sussex, England BN7 1XU

Distributed in Australia by Capricorn Link (Australia) Pty Ltd.,
P.O. Box 704, Windsor, NSW 2756 Australia

If you have questions or comments about this book, please contact:
Lark Books
67 Broadway
Asheville, NC 28801
828-253-0467

Manufactured in Canada

ISBN 13: 978-1-60059-490-8

For information about custom editions, special sales, premium and corporate purchases, please contact Sterling Special Sales Department at 800-805-5489 or specialsales@sterlingpub.com.

This book was printed on 100% post-consumer recycled paper with agri-based inks.

♥ Table of Contents

Introduction

You may know of—or know personally—folks turning the basements of their brownstones into root cellars. Or people putting chicken coops in their suburban backyards. Or families experimenting with making their own mozzarella or buying berries at the local farmers' market and giving canning a try. Maybe you're one of them. Maybe you want to be. Or maybe your interest has simply been piqued by the renewed interest in local foods and how they're grown, raised, and made. Whatever the reason, welcome to the adventure, the possibilities, the ongoing experiment, and the sheer joy of crafting a homemade life.

This book aims to give you clear information plus tons of inspiration, whether you're ready to start building your coop or just curious about what's involved in caring for a small flock of your own.

Chickens may bring to mind visions of acres of land and big red barns, but, truth be told, they're the easiest and most rewarding creatures for a fledgling urban, suburban, or rural homesteader to manage. They're happy to call all kinds of structures home—from traditional wooden and screen housing to tarped wooden pallets and even dog houses—and they're a natural fit with children. On top of all that, they'll provide you with an ongoing food supply that's fresh and as local as it gets.

I opened wide the door to my own personal chicken journey one fateful spring afternoon. With optimism and excitement, I reserved a flock of five newborn chicks and got busy setting up house. It wasn't until I was driving down a steep, deeply pocked mountain road, my newly acquired wards murmuring various degrees of discontent from the dog crate in the rear, that I realized I had an awful lot to learn about chickens.

I decided to start keeping chickens for two reasons: pragmatism and pleasure. Keeping laying hens allows me to craft a nutritious meal merely by popping into the backyard and gathering the eggs warm from the nest for breakfast, lunch, or dinner. What could be more local than foods gathered from your own backyard, stoop, balcony, or deck? The pleasure component comes from the chickens themselves. Simply put, they are imminently captivating. The ceaseless squawking, pecking, digging, eating, running, preening, laying, and roosting provide fodder for countless hours of entertainment.

No matter the size of the place you call home, it's pretty likely you'll be able to fit some chickens in there. You really don't need much room to keep a few chickens; these 4- to 8-pound (1.8 to 3.6 kg) creatures have no hooves or snouts, just beaks and wings. If your space prohibits keeping poultry of any kind, there may be a chicken-keeping neighbor or chicken co-op nearby willing to let you join in on the fun.

Keeping chickens can also give you the opportunity to network in your community. Perhaps you'll make connections while soliciting advice online. Maybe you sell your eggs at your local farmers' market and connect with the folks you feed on a regular basis. Or it could be simply picking up chicken mash at the feed store or chatting online with other "chicken tenders" that brings you together with people you might have otherwise never encountered.

In this book, I share with you all the nitty-gritty details and learned-from-experience tips discovered along my own chicken voyage. I've described the basic steps needed to get you started off right from the first day, from deciding what breed is best for your particular needs to feeding, cleaning, housing, and caring for your flock. You'll find plans to make your own nesting boxes; a real dynamo of a moveable chicken tractor; delicious egg-centric recipes created, tested, and consumed in my own kitchen; and plenty of humor and camaraderie. My sincere hope is that this reference will serve as a useful and enjoyable companion to you along the way, giving you the guidance I wish I'd had when I got started. Take the plunge into this new way of living, and your life will open up to embrace the wondrous magic in ordinary things. It's messy and exhilarating ...and I think you'll love it as much as I do!

Ashley English

ABOUT THE AUTHOR

A few years ago, I was hopping into my car each morning, heading off to a job in a medical office. Things changed, though, when a whirlwind romance quickly resulted in marriage, a little homestead at the end of a dirt road, and just the encouragement and support I needed to make some serious life changes. Combining my long-standing interest and education in nutrition, sustainability, and local food, I made the bold decision to leave my stable office job and try my hand at homesteading. It was a huge leap of faith, but I truly believed there was opportunity waiting in a simpler, pared-down life. My goal was to find ways to nourish both body and soul through mindful food practices. And so I jumped in, rubber boots first, completely unaware of what lay ahead.

In my desire to chronicle both the triumphs and lessons of crafting a homemade life, I started up a blog, Small Measure (small-measure.blogspot.com). In it, I aspire to convey the same ideals I live every day: There are small, simple measures you can take to enhance your life while also caring for your family, community, and the larger world. It's been a trial-and-error experiment in living, full of a few pitfalls along with the joy. I've learned so much along the way, and I hope that this book serves as continual encouragement for you. If I did it, you certainly can, too.

Chapter 1
What to Consider

Welcome to the wonderful world of chickens! Whether you want your own fresh eggs each morning, desire a trusted, wholesome source of meat, or simply wish to add a new pet to your home, chicken raising offers a multitude of rewards. However, before you crack the first catalog or get misty eyed over some antique feeder, you must address several essential preliminaries. Determining if you have adequate time to care for your flock, whether or not you can even keep chickens where you live, and what your neighbors think about having poultry next door are a few of the concerns to address before you get your heart set on that gorgeous Buff Orpington with the adorable pantaloons.

TIME AFTER TIME

Perhaps the first thing to consider when entertaining the idea of backyard animal husbandry is that caring for chickens takes up time. The amount of time will vary based on their age as well as the size of your flock. If you opt to start out with chicks, be prepared to allocate several hours each day to babysitting them, making sure they haven't harmed themselves or each other, and that all their physical needs are met. Baby chicks need to be checked several times throughout the day.

If your birds are older, they will still need to be fed, cleaned up after, and, depending on their housing situation, let out in the morning and locked up safely away from predators at night. Eggs should also be gathered several times daily.

MONEY MATTERS

In addition to taking up your time, keeping chickens will also take up some of your money. Start-up costs will be the bulk of your chicken-raising expenses. At minimum, you should plan to spend several hundred dollars for purchasing or building a coop, complete with nesting boxes and perches; acquiring feeders, waterers, and feed; and buying the birds themselves. Your setup can be as simple or elaborate as your budget allows. Thrifty, enterprising individuals can craft a coop from a truck cab or a few 2x4s and some chicken wire. Chicken aesthetes might opt for a modern mobile unit or a chicken ark. We'll cover housing in detail in chapter 4, but be mindful for now that it is important to determine if you can afford to develop your chicken infrastructure. Once your chicken compound is all in place, future costs are generally limited to feed, purchasing new birds, and veterinary fees and medications, should either be needed. As you begin thinking about raising chickens, take the time to run the numbers and make sure you can manage the necessary start-up investment and ongoing costs.

WEATHER OR NOT

Selecting chickens suited to your climate and varying weather conditions is another important consideration. One chicken's balmy paradise is another chicken's hothouse of death. Like- wise, a fluffy-feathered Brahma will fair well through frosty New England winters, while a Mediterranean breed such as a Leghorn might have a rough go of it if not properly housed.

As you begin toying with the idea of owning and caring for chickens, don't forget to do a little research on their climate needs. Doing so will go a long way toward preventing unneces- sary stress for the both of you. A listing of breeds and climate hardiness can be found on page 33.

SPACE IS THE PLACE

While chickens don't need much space, they do need some. I've seen chickens kept in the 8-x-10-foot (2.4-x-3 m) cement backyard of a Scottish bed-and-breakfast, along with rabbits, goats, pigs, and other animals. At my house, the Ladies have an inordinately large coop and run, owing to the coop's previous incarnation as a dog shed for two immense German Shepherds.

Your particular needs will be based on the size of your flock, the size of your breeds, and the amount of outdoor space your chickens will be allocated, if any. In general, chickens need 4 square feet (1.2 m²) per bird if given access to outdoor runs and 10 square feet (3 m²) per bird if they have no outdoor access. That amount will be halved if your birds are bantams. More about space considerations will be discussed in chapter 4 on housing.

It is important to evaluate what space you have available before you begin purchasing birds. Too little space and you can have ornery chickens. Too much and you run the risk of poten- tially annoying your neighbors and making more cleanup for yourself than necessary.

BEING NEIGHBORLY

After determining if you have the time and space to devote to your feathered friends, your next step should be checking with your neighbors to learn their position on the notion. This is especially important if you live in close proximity to other homes or apartments. Discuss your plans, including the size of your flock, the housing you intend to construct, and the care and maintenance you have in mind. Promises of eggs left on their front stoop (by you, not by the chickens themselves!) may not be enough for some folks. Letting them know there won't be any roosters present as well as your plans for housing and clean-up could assuage the concerns of those not completely bowled over. Doing so in advance of the actual physical appearance of your flock can save headaches and potentially money, should your chickens offend and your neighbors pursue legal action. If you live more than a quarter mile from your closest neighbor, checking in with them may be polite, but most likely unnecessary.

One of my favorite unexpected perks in keeping chickens is the daily lessons they offer in mindfulness. Chickens live in the moment, thrilling at the conquest of a wriggling grub, squawking in triumph at the delivery of an egg, resting contentedly in a dust bath. They don't worry about whether they spent too much time in that dust bath, or if they squawked too loudly about that egg, or if they ought to have squirreled away that grub for another day. They rise with the sun and get to the business of living with a vivaciousness, curiosity, and deliberation we could all learn from. While you may be setting out on your own chicken-raising adventure seeking nourishment for your body, I predict you just might find some for your soul, too.

CRACKING THE CODE

After clearing your plans with your neighbors, the next essential step is to determine whether keeping chickens is permitted where you live. Variations in codes and ordinances exist from state to state, as well as between cities. If you live in a rural area, keeping chickens is almost assuredly legal. However, if you live in a city or suburb, it is imperative to find out what's on the books about raising chickens legally.

This can be done through several routes. Search online or in the phone book for your local county or city commissioner's office. Look for specifics on animal control rules. If an online inquiry fails to locate information on municipal codes for keeping chickens in your area, a direct call to your local municipal animal control office will get you to someone who knows. The local library is also a good place to search for information on local animal ordinances. Some cities will allow a certain number of chickens to be kept without a permit, but flocks in excess of the allotted number require a paper trail. Most permits can be had for a nominal fee and need renewing annually. Keeping the permit is most often conditional upon a lack of complaints from neighbors. Remember, check with the neighbors first! Depending on your local ordinances, it may be necessary to first obtain written consent from your neighbors in order to get a permit or keep chickens at all.

Furthermore, depending on the size of your intended coop, you may need a building permit as well, although more than likely a permit will not be necessary. Check with your local government planning and development office to be sure.

SOMETHING TO CROW ABOUT

Keeping roosters is prohibited in most urban and suburban municipalities. Assuring your neighbors they won't be woken by crowing at the crack of dawn might even be essential to obtaining their consent. While roosters may be majestic to look at, they aren't an essential part of a backyard flock. Hens will lay eggs regardless of whether a rooster is present or not.

Roosters are required for breeding purposes, though. If you live in an area where keeping roosters is not allowed, you will need to replenish your flock another way. Breeding and care for chicks and eggs will be discussed in chapters 6 and 7.

Roosters, while a joy in their own right, may not always be welcome in your neighborhood.

When allowed, keeping a rooster can be a fantastic way to enable breeding chicks at home. Watching a rooster proudly strut his stuff and enliven the airwaves with his crowing can be sheer joy to witness. Roosters also help maintain a pecking order, although one will be established without a rooster around. In my flock, Georgette, the Barred Rock, rules the roost. Even though I live in the country, I've opted not to have a rooster, and the Ladies enjoy a celibate sisterhood in a sort of hen convent. They couldn't be more content.

CAN'T WE ALL JUST GET ALONG?

If you have other animals, it is important to consider how they may interact with your flock. Barnyard cats and adult chickens generally can achieve a peaceful coexistence. Chicks and cats are a lethal combination, though, and caution should always be exercised to keep cats away from chicks until they are of approximately equal size.

Dogs, however, are an entirely different matter. Some dogs will show complete disinterest in your chickens, while others can only be trusted near the chickens with human supervision. My wonderful, kind, gentle, 80-pound (36 kg) German Shepherd, Fly, went nuts with bloodlust when I first brought the chicks home. To curb her behavior, I began a slow, consistent routine of having her come with me to the coop each day, observing as I fed the chickens, stroked them, gathered eggs, and generally socialized with them. Fly is considerably calmer around them these days, but she still likes to jump up on the fence as we approach the coop.

Regardless of where you live, in the country or the city, be aware that allowing chickens to range freely without fencing may make them targets for wandering dogs. I've heard of many folks who lost a hen or two, or 13, to marauding dogs. Gauge your own dog's interest, and note the likelihood of wandering dogs as you consider your chickens' accommodations and fencing.

CHICKENS AND CHILDREN

Some wonderful magic happens between children and chickens. Kids just can't seem to get enough feathery fun. If you have children and are considering raising chickens, a bit of caution is advised, however. Some birds can be aggressive, while others are prone to flighty behavior or nervousness. Select a breed known for a docile, friendly disposition when children will be interacting with them regularly.

Furthermore, proper hygiene will need to be exercised when children handle chickens. Chickens are generally in pretty close contact with their own poop. In order to access their coop, it is often necessary to wade through at least some poop. For these reasons and more, children should wash hands thoroughly after handling chickens.

You might also want to consider shoes reserved just for forays into the chickens' housing. I have a pair of rubber boots whose express purpose is to be worn in the coop. I don my boots, grab the chickens' food, and yell out "Let's go get the chickies!" to the dogs. Once we return from feeding the chickens or gathering up eggs, I remove the boots and park them immediately next to the door. Kids would get a kick out of having "chicken shoes." You would get a kick out of not having chicken poop dispersed throughout your home.

Portrait of a chicken owner

Theresa

When she isn't busy drawing blood or hooking up IVs to her human patients, Theresa, a registered nurse, skillfully tends to the needs of her small flock of five. She sourced her flock, comprised of three hens and two copacetic roosters, from a local feed store.

Childhood visits with her grandparents and their menagerie of animals, which included chickens, are treasured memories for her. When it was time to keep a flock of her own, Theresa had to consider the particular needs of her environment. She lives on a wooded mountainside, where any number of threats to her birds abound. Roaming neighborhood dogs, coupled with a forest full of predators, including wild coyotes, made free ranging out of the question.

In order to keep her feathered friends safe, Theresa and her husband fashioned a chicken tractor out of on-hand materials. Like Theresa, be sure to consider the predators in your area posing a threat to your flock, and create housing appropriate to those needs. Also, it's essential to take the time to evaluate the number of chickens you really want or need to keep before bringing home the first tiny peeper. As Theresa advises, **"be realistic about your daily egg needs."** Her ex-husband once enthusiastically ordered 70 chicks from a hatchery, resulting in far too many eggs for a household of two.

Permaculture

Permaculture is a combination of the words permanent and agriculture, as well as permanent culture. Australians Bill Mollison and David Holmgren, attempting to describe an approach to creating lasting agricultural systems that mimic nature, coined the word during the 1970s. Permaculture principles borrow greatly from those that govern natural ecology, wherein organisms synergistically complement one another. When applying permaculture tenets to design, the particular needs, properties, and potential output of a physical space are constructed in relation to one another.

Applying permaculture principles to raising chickens involves considering then what a chicken needs, what it produces, and what its inherent properties are. All chickens need water, food, other chickens, and specific climatic conditions. When those needs are met, they produce eggs, meat, fertilizer, and feathers, in addition to breaking up and enriching the soil and eating weeds and unwanted insects.

When incorporating permaculture into a plan for raising chickens, all elements of the design are created in relation to one another, "stacking functions" as it were, so that each design element reinforces, nurtures, and complements the other. Physically, this may manifest as a chicken run that is planted with perennial trees and bushes from which the chickens may forage seeds and fruit themselves. Certain times of the year may necessitate the addition of nutritional supplements, but in a well-designed permaculture setup, the chickens will be able to meet many of their food needs themselves. Set a rain barrel atop the chicken coop, and much of the chickens' water needs can be met with minimal energy expenditure. For housing, a greenhouse could be built along a south-facing wall, providing heat for chickens in the winter and carbon dioxide from the chickens' breath for plants, to help them grow.

Portrait of a chicken owner

Patrick and Holly juggle busy careers as graphic designers with their roles as coordinators of an all-volunteer egg cooperative in Portland, Oregon. The collective, known as the Eastside Egg Co-operative, was partially founded by a local CSA (community supported agriculture) farm and by Heifer International, a nonprofit organization committed to ending world hunger and poverty by providing livestock, plants, and sustainable farming education to poor families around the world. Co-op volunteers hold weekly shifts, either rising with the chickens or putting them to bed, so that the flock is cared for twice daily. In addition to gathering up eggs and providing feed and water, Patrick and Holly work to ensure volunteers provide minor maintenance and upkeep to the flock's housing.

The 40 hens are kept in a moveable lightweight structure fashioned out of two bicycle wheels, custom-made tubular steel forks (both sourced from local suppliers), and a wooden frame. Chicks being raised to replenish the laying flock are kept in a separate pullet house. The birds live mostly outdoors, corralled in by electric netting, where they are rotated among 12 small fields.

Although they didn't grow up with chickens, Patrick and Holly have taken to chicken raising with laudable dedication and tenacity. They were prompted by a desire to educate themselves and the wider community about animal husbandry practices, as well as an interest in the promotion of urban agriculture in general.

Holly

Patrick

When selecting a breed, Patrick has a bit of advice for those living in areas where the mercury drops below freezing. **"We generally recommend that backyard poultry keepers choose a heavy layer breed such as the Barred Rock, Orpington, or Rhode Island Red. These breeds are agreeable and easy to work with, and they withstand cold better than smaller breeds. Most of them are dual-purpose breeds, meaning that they will make fine stewing hens when the time comes, as well."** Furthermore, if you are mixing breeds, you can expect feathers to fly on occasion. **"They're going to fight. It's okay. Worry only if there is blood,"** advises Patrick.

Chapter 2
Selecting a Breed

Choosing the right type of chicken is a bit like choosing a mate. Generally, we look for an individual whose interests mesh with our own. Does he like Woody Allen films? Is she fond of Mark Rothko? Can he savor and appreciate a fine smoked Gouda? Of course, questions of marriage, children, career, and such must also be addressed. Chicken selection is not entirely dissimilar. It is important to consider your particular needs and desired characteristics when picking your flock. Careful selection will help to insure a mutually satisfying relationship.

A breed is a group of chickens, most likely distant relations from the same stock, sharing similar characteristics, including size, skin color, and comb and plumage styles. Within their breed, chickens may then be further subdivided into varieties, where distinguishing colors and patterns differentiate, say, a Silver Laced Wyandotte from a Columbian Wyandotte.

There are over 200 breeds of chickens out there, ranging in size from 1 to 12 pounds (.5 to 5.5 kg), and, on rare occasions, larger. Breeds vary in temperament and climate needs, among other distinctions. Take your time, consider your preferences, and choose wisely. As with any good relationship, take into account the needs of both you and your partner; in this case, your partner being a feathered, winged, squawking chicken.

Keep in mind that, no matter your situation, there is a chicken for every scenario, whether the place you hang your hat is a palace or a mere perch. Consider your particular needs, jot them down if needed, and then begin your hunt. Scour catalogs, visit feed stores, and peruse chicken websites with fervor, holding out until you find Mr. or Ms. Right.

THE MAGIC NUMBER

Depending on the size of your property and the size of the human flock that will be consuming what your chickens are offering, you may need few or many. Will you be keeping eggs exclusively for your family, or for market? Will you be keeping chickens for meat or for eggs, or for both? Will your flock be for show or for production? Will your chickens live in a tractor or a coop? Will you keep a rooster? Do you have an acre or an alleyway?

The average backyard chicken keeper may maintain a flock of as few as two hens, or upwards of 100 birds, depending on their needs. No matter what number you ultimately decide on, bear in mind that chickens are naturally social creatures. A minimum of two will keep their spirits up, as well as provide even more chicken antics for you to relay to family and friends, or to total strangers, as I sometimes do!

SIZE MATTERS

Size, perhaps more than any other characteristic, is what truly differentiates chicken breeds from one another. Size considerations can be paramount. Again, consider the space available for your flock. Is it more suited for bantams, which are small breeds, or for standard breeds?

Bantam Island, located in the Dutch East Indies, is a seaport once heavily trafficked by European sailors. They often selected the small fowl native to the area for long-haul trips, owing to their compact stature, usually somewhere between 2 and 4 pounds (1 to 1.8 kg). The term "bantam" came to be applied to any small chicken breed, whether from Bantam Island or not. Many bantams have a standard-size counterpart, although several, including Silkies and Bearded d'Uccles, are available only as bantams.

While many bantams are kept purely as show birds, all bantam hens will lay eggs, albeit small ones. (For cooking, in recipes where you might use two large eggs, you'll need three bantam eggs.) Although their output in no way rivals that of standard-size breeds, bantams may be the ideal choice for the would-be chicken keeper with space restrictions. Bantams are typically less than half the size of standard varieties.

These petite birds have correspondingly small appetites, which cuts down on feed costs for their keepers. Bantams love to graze, hunt, peck, dig, and generally happily seek out and annihilate any creepy crawlies, further diminishing their need for feed. Small size and little appetite may be the perfect combination for those keeping an urban flock on a ramen-noodle budget. Do keep in mind, though, that bantams are flighty, with the ability to reach up to 6 feet (1.8 m), and will need to be fenced accordingly.

Brahma
Jersey Giant
Bantam

PURE OR HYBRID?

Deciding on your flock with purebred status in mind may or may not be important. If you someday wish to show your chickens, then they must be purebred. Just as mutt dogs, no matter how flawless and adorable, will never be put on parade at the Westminster Kennel Club, hybrid birds will never make it on the chicken show circuit. It's a harsh, cruel world, I know. Aside from the sheer fun of raising purebred chickens, with all their wild variation and general sauciness, sticking with recognized breeds is a wonderful way to connect with chicken fanciers and breeders.

Even if you have no intention of ever putting Henny Penny on the catwalk, whatever supermodel tendencies she may exhibit, you may wish to keep and preserve a rare or endangered breed. Continuance of numerous so-called "heritage" breeds such as Araucanas, Dominiques, Dorkings, Spanish, Buckeyes, and Aseels rests in large part on small-scale chicken owners. As interest in backyard chicken raising grows—recent trends point to a nation consumed with chicken fever—perhaps these heritage breeds will cease to be endangered.

Regardless of the many appealing characteristics of purebred poultry, in no way should hybrids be considered inferior. Hybrid chickens are the result of genetically crossing two distinct breeds. While they may lack the silver-spoon pedigree of their purebred cousins, they excel in other ways. They have been bred to possess the best possible traits of the breeds they were crossed from. Hybrids may fall short on some criteria as articulated in the Standard of Perfection, such as length of tail feather or pattern of stripes on their neck, while embodying other desirable attributes, such as generous egg or meat production. Hybrid vigor also makes crosses especially sturdy and resistant to disease. Do bear in mind that hybrid birds will not necessarily bear chicks in their own likeness. In order to obtain a new generation of a particular breed, you will have to go back to original stock.

LAY, LADY, LAY

Many people elect to raise chickens for the singular thrill of gathering up still-warm eggs. Etched forever in my mind will be the first time I tasted, and viewed, eggs from my own chickens. I never knew eggs could be so delicious, so creamy, so, well,

orange! When my husband, Glenn, began whisking up the first offerings from our girls, I was convinced he must have added a pinch of turmeric to the bowl to produce that intense hue. He hadn't—the yolks were just that vibrant, enhanced by all the parsley, salad greens, and berries I had been adding to the Ladies' feed. These foods all contain carotenoids, nutrient-rich pigments praised for their immune system-enhancing properties. Supplementing chicken feed with such foods results in almost pumpkin-colored yolks, transferring all that carotenoid goodness to whomever we fire up the skillet and break our eggs for.

If your primary motivation for raising chickens is to be provided with a continuous supply of eggs, consider a breed known for its heavy laying abilities. Of course, all hens will lay eggs; some breeds are merely known for their inclination to lay with a greater degree of regularity. The breed best known for its profusion of eggs is the White Leghorn. Other top laying breeds include Minorcas and Rhode Island Reds.

Egg production will be most abundant during a hen's first two years, after which her laying will become a bit more sporadic. Depending on your needs, this tapering off may work just fine, or you may consider introducing new chicks or pullets to your flock over time to guarantee a continuous supply. Assuming average consumption and production, two hens for each family member should take care of your egg needs. Refer to the chart on page 32 for a listing of breeds suitable for those with a hankering for eggs.

SETTING THE TABLE

Perhaps your interest in raising chickens lies in their ability to keep your table full of safe, clean meat you can feel good about consuming. Factory-raised chickens are exposed not only to antibiotics, but often to cruel and inhumane treatment. Birds raised on your own plot, fed by your own hand, offer peace of mind alongside a source of healthy nourishment.

As much as you may care for their well-being, table birds are not pets. Do not allow yourself (or your children, should you have any) to become too attached to chickens intended for the table. Fortunately, table birds are ready for butchering at about six to eight weeks. Such brevity of time on earth makes it difficult to foster too deep a tie with your meat chickens.

The breed most commonly used for table is the Rock-Cornish or Cornish-Rock Cross, a hybrid of the White Rock and Cornish Game. Large commercial meat bird factories favor this breed as it develops quickly and has a good meat-to-bone ratio. With voracious appetites and little interest in getting much exercise, Rock-Cornish Crosses can gain 4 pounds (1.8 kg) in eight weeks. Their rapid growth rate may contribute to health and structural problems, and therefore the need for careful

handling. If you do opt for a hybrid table breed, be aware that such birds will not breed true. Should you be interested in breeding more, you will need to return to original stock for desired characteristics.

You can also opt for purebred chickens to use as table birds. Table breeds other than the Rock-Cornish Cross include Jersey Giant, Cochin, and Brahma. Although these birds mature at a slower rate before they are ready for table, their slower growth rate prevents many of the problems affecting the rapidly growing hybrid. As they are generally large and heavy, many table birds will not require much fencing, which may be ideal for those with limited space or funds.

In the interest of full disclosure, I will confess that I don't eat meat. That said, I fully support the decision of anyone—city dweller, suburbanite, or countryside resident—to raise chickens for consumption. I firmly believe it to be inherently empowering and sustainable.

BUTCHERING

Birds intended for table should be butchered around 6 to 8 pounds (2.5–3.5 kg). Do this in the evening, when your flock is calm and relaxed. Butchering chickens, while not difficult, is best performed after witnessing, perhaps repeatedly, someone else do it who has done it before. Perhaps a family member, a neighbor, or a fellow chicken raiser in your community can show you the ropes. If you feel that this is a task you might never be able to perform, there may be a local butcher or mobile poultry processor nearby willing to perform the job at a cost. Ask around at the feed store or solicit a recommendation from your veterinarian.

TWICE AS NICE

Should you be interested in chickens for both eggs and meat, seek out dual-purpose birds. These birds will reliably provide you with an adequate egg supply and are known to be good to eat. Dual-purpose breeds lay fewer eggs than those known for their egg-laying abilities (about 18 to 20 dozen in a laying cycle as opposed to 20 to 22 dozen) and will put on weight much more slowly than birds raised primarily for table, but are more adept in other ways.

The criteria used in judging at poultry shows in the U.S. are detailed at length in the Standard of Perfection, published by the American Poultry Association. If you want to know desired breed characteristics for some form of poultry, be it chickens, ducks, geese, turkeys, or bantams, this is your go-to book. Outside the U.S., nation-specific rules will apply. An online search should provide the necessary information for judging breeds internationally.

Keeping chickens to show can be great fun, although it requires a bit more elbow grease on your part. Ever given a chicken a bath? Now's your chance! Chicken shows also enable chicken fanciers an opportunity to meet and greet. I love that term, "chicken fancier." I can almost imagine my third-grade teacher Mrs. Pierson asking "And what do you want to be when you grow up, Ashley?" "A chicken fancier!" I wonder if "chicken fancier" is a vocation listed on the census?

My personal favorite show bird is the Yokohama. The tail length of a mature Yokohama is two feet! Fanciers in Japan often keep Yokohamas in conditions preventing molt, thereby allowing their tails to grow to up to three feet. Such proud plumage I imagine must make even a peacock jealous!

If you are considering keeping chickens for show, look for small poultry societies in your area. The American Poultry Association or your local extension office would be good places to seek out these small organizations.

Shows are customarily held in winter, by which time spring-hatched chicks will be fully grown and older birds will have new plumage post-molt. What better way to brighten up the winter doldrums than by gussying up your favorite bird and mingling feathers with fanciers and their fine feathered friends!

Dual-purpose breeds are generally hardier and more self-reliant than single-purpose breeds. Many are broody, meaning they will sit on a group of eggs (a "clutch" in chicken terminology) to hatch. Broodiness can be beneficial in chickens, whereas broodiness in teenagers can result in such unpleasantness as petulant whining and refusal to make eye contact.

Having been less-intensively selected for specific, narrow characteristics, most dual-purpose birds are also generally calmer than more specialized breeds, making them an ideal choice for the beginner. They are less flighty and will therefore not require high fencing. Dual-purpose birds may begin laying a bit later than other breeds, at around five-and-a-half to six months instead of five months. As for their meat, these breeds purportedly produce more tender meat than exclusively egg-laying breeds.

SHOW ME THE CHICKENS!

Some chickens are kept exclusively for purposes of show. Coddled, preened, and stroked their entire lives, show chickens lead cushy lives anyone might covet. They are hand-groomed, often live in their own separate housing, and enjoy no small amount of attention and, perhaps, fame!

Chicken Anatomy 101

points

blade

comb

base of comb

eye

ear

beak

sickles

saddle

ear lobe

wattle

back

cape

hackles

breast

secondaries
(wing-bay)

main tail

saddle feathers

vent

primaries
(flight feathers)

shank

thigh

spur (males only)

toenail

BREEDS

More than 60 breeds of chicken are recognized by the American Poultry Association. Each can be distinguished by such characteristics as size, plumage coloration patterns, comb type, and personality. The following breeds provide a sampling of varieties available to the backyard grower. Those listed include breeds known for being good layers, suitable for table, dual-purpose, or for show. Information on additional breeds can be found online or in poultry catalogs.

Feather Descriptions

Barred	Alternating transverse markings of two distinct colors
Columbian	Typically white plumage with a ring of black feathers around the neck and black tips on the ends of the tail
Laced	A border of contrasting colors around the web of the feathers
Penciled	Most commonly concentric rings of alternating colors, although other patterns do exist
Spangled	A distinctly colored V-shaped pattern on feather tips

Araucana

This breed is often referred to as the "Easter Egg" chicken, as its eggs are a pastel green-blue color. Araucanas originated in Chile and are also known as the South American Rumpless. There is some speculation that the Araucana may have initially been brought to South America by Polynesian sailors visiting the country before Columbus, bringing the ancestors of today's Araucanas with them. Distinguishing characteristics include the lack of a tail, tufts behind the ears (giving them a resemblance to a chicken Albert Einstein), and their egg color, which Ameraucanas and Easter Eggers also produce. The breed offers up to twelve recognized plumage color variations, depending on whether the judging is based on American or British criteria. The average hen size is 5½ pounds (2.5 kg), which would make it a good candidate for those desiring chickens of a smaller size. Classified as light, meaning their ancestors are of Mediterranean or Asiatic origin, they are known as good layers.

Brahma

Brahmas hail from the Brahmaputra region of India. They are believed to be related to the Junglefowl and the Cochin. Often called the "king of chickens," Brahmas are especially large, with roosters weighing in at 11 pounds (5 kg) and up. Recognized varieties in the American Standard of Perfection include light, dark, and buff. Like Cochins, Brahmas are heavily feathered, including their feet and legs, and will need to be kept out of bad weather to avoid sullying their lower feathers.

Brahmas are considered a slow-growing table bird, taking up to two years to reach full development, making them less attractive to commercial chicken growers. Many fanciers keep them primarily for show purposes, or simply as pets. Brahmas lay relatively small eggs. They are relaxed, calm, placid birds, not predisposed to flightiness. If kept in a mixed breed environment their calm demeanor can make them targets, and you will need to keep a watchful eye to prevent bullying.

Cochin

Developed in China, this breed was originally known as the Chinese Shanghai. It was brought to the UK and United States in the 19th century. These are especially large birds, with roosters weighing in around 11 pounds (5 kg). Much like the slow-maturing Jersey Giant, Cochins take time to develop for table readiness. They are reasonably good layers, and are known for their broodiness.

The Cochin's most distinctive feature is its abundant plumage, covering even its legs and feet. As such, this breed will need to be kept inside during inclement weather to avoid ruining its lower feathers. Cochins possess a small single comb that may on occasion be serrated. Thirteen varieties are recognized by the APA including partridge, barred, golden laced, mottled, silver laced, and birchen. Their size enables them to take well to cold weather, and they are known as friendly, easily handled birds.

Frizzle

Kept primarily for show, Frizzles are distinguished by their feathers, which curl outward, as opposed to lying flat as on most chickens. Asiatic in origin, Frizzles originally appeared 300 years ago in southern Asia, Java, and the Philippines. In some countries, any bird exhibiting this sort of plumage is considered a Frizzle, but in the United States and the United Kingdom it is considered a distinct breed unto itself. This breed appears in a range of colors, including blue, black, white, buff, and silver-gray. Colors patterns also allowed for show but showing up less often include Columbian, duckwing, black-red, brown-red, cuckoo, pile, and spangle.

Frizzles are fair egg layers, offering light brown and white eggs. Furthermore, they are regarded as excellent table birds. They are known as brooders, although to successfully breed chicks with plentiful and sturdy plumage, use a young rooster to breed with your pullets. On account of their unusual feather structure, they are not well suited to wet climates and would be best kept indoors. Until fairly recently, most Frizzles kept by chicken enthusiasts were bantams. A breeding program to revive large sizes has proven successful, although both bantam and large Frizzles are still considered rare.

Jersey Giant

Developed in New Jersey, this breed weighs in at 9 pounds (4 kg) and up, classifying it as a very heavy breed. Jersey Giants were created by crossbreeding Orpingtons, Black Javas, Dark Brahmas, and Black Langshans. Previously, roosters grew up to 20 pounds (9 kg), although today's varieties are nowhere near as large. While the Jersey Giant will take longer to mature for table than some hybridized breeds (around six months as opposed to five weeks), it is recognized as a good table bird.

Given its size, it will need a good amount of space, and since it matures slowly, it will consume a considerable amount of feed before becoming table ready, considerations to bear in mind when looking at your size and budget limitations. Jersey Giants are known as easygoing, docile birds, which will fare well in cold weather on account of their size. They are not known as good brooders. Possessing a single comb, Jersey Giants are available in three varieties: white, black, and blue. To date, they remain the largest breed of chicken developed in the United States.

Leghorn

When the average person hears the word "chicken," an image of the Leghorn will likely come to mind. Anyone familiar with the Looney Tunes cartoons will remember the oversized, Virginia-accented, perpetually hen-wooing rooster known as Foghorn Leghorn. Named after Livonro, the Italian city from where the breed originated, Leghorns are mostly found with white plumage, although 11 other color variations including red, black, blue, buff, Columbian, buff Columbian, barred, and silver varieties are also recognized by the American Bantam Association and the American Poultry Association.

Leghorns are renowned for their egg-laying abundance, around 300 annually. For this reason the White Leghorn is one of the most common commercial breeds in the U.S., and most of the eggs sold in grocery stores are of the Leghorn variety. Leghorns all possess single combs, and some possess rose combs, which, though rare, are allowed for exhibition. Full-sized, the birds average 4½ pounds (2 kg) and are considered flighty. Their combs need to be protected from frostbite in the winter.

Orpington

Englishman William Cook developed the Orpington in 1886. Combining Plymouth Rocks, Langshans, and Minorcas, he named the new hybridized breed after his hometown. Orpington colors include black, white, buff, blue, and red and have single combs, although Black Orpingtons may have a rose comb. It is considered a heavy breed, weighing between 7 and 10 pounds (3 and 4.5 kg), and its abundant plumage makes it appear even larger.

As the Orpington has a reputation for being an especially docile bird, it is an ideal choice for those just beginning to keep chickens. For this reason they are also a good choice for families with small children. Orpingtons are good layers, averaging 110 to 160 eggs annually, and they continue to lay year-round. Their eggs range in color from light brown to off-white, varying in size and color based on hereditary traits.

Plymouth Rock

Plymouth Rock is an American breed, more commonly referred to as a Barred Rock on account of its plumage pattern, which displays alternating "bars" of color, most commonly black and white. In addition to the barred variety, recognized varieties of Plymouth Rock include blue, buff, Columbian, partridge, silver penciled, and white. The White Plymouth Rock female is bred with the male Cornish Game to create the most popular modern-day broiler, the Cornish-Rock cross.

Plymouth Rocks are dual-purpose fowl, suitable for both table and egg production. Until World War II, Barred Rocks were the most common chicken breed in the United States, stemming in no small part from the breed's known traits of calmness, broodiness, and cold-hardiness. Plymouth Rocks lay large brown eggs, averaging around 200 annually. They are considered a heavy breed, at 7 to 8 pounds (3 to 3.5 kg), and possess a single comb.

Polish

What is the craziest looking chicken you ever saw? The Polish chicken, that's what! Resting on a cone atop their skulls (referred to as a "protuberance") are huge, puffy, ball-like crests of feathers. In addition to a small V-shaped comb, the crests cover their heads completely. Typical of small breeds (hens are 4½ pounds [2 kg]; roosters, 6 pounds [2.7 kg]), the Polish has large nostrils. A European breed, Polishes did not actually originate in Poland, as their name would imply, but instead are suspected to be from the Netherlands.

As they are now bred primarily for show, Polish are not broody. They are fair layers of small white eggs. Their crests limit their vision, and although friendly, this vision impairment can make them a bit skittish. They are most often found at the low end of the pecking order when housed with other breeds. Care will need to be taken to keep the crests in good condition, mostly covered from inclement weather. Ideal as conversation starters, Polish would make a great addition to any backyard flock.

Rhode Island Red

As its name suggests, the Rhode Island Red was originally developed in Adamsville, Rhode Island. In fact, they are the state bird of Rhode Island today. Their vibrant red plumage is the result of interbreeding with the Malay breed, resulting in a rust color, although darker shades bordering on maroon exist. The comb is most commonly single, although a rose comb is not unusual. They possess a strong constitution, foraging heartily and not succumbing easily to illness.

Rhode Island Reds are prized for both their meat and eggs. Excellent egg layers, they can average 250 to 300 eggs annually. They are known as cold hardy, although their combs are susceptible to frostbite in particularly cold climates. While generally friendly, they can become nervous and possibly aggressive around unfamiliar faces, a good thing to keep in mind with small children. Repeated exposure will calm them, though. Although they are considered a separate breed, Rhode Island Whites share many of the same characteristics.

Wyandotte

The Wyandotte is a heavy bird, averaging 7 to 8 pounds (3 to 3.5 kg). Originating in the United States, possibly New York, Wyandottes are distinguished by a laced marking on their plumage. This patterning is believed to have been developed by crossing a Silver Sebright Bantam with either a Cochin or Brahma, then later with a Spangled Hamburg. Officially recognized varieties include silver laced, golden laced, white, black, buff, partridge, silver penciled, Columbian, and blue; the blue-laced red and barred can also be found, although they are not recognized by the APA.

Wyandottes are considered dual-purpose birds, maturing slowly for table and laying large brown eggs, averaging 200 annually. In addition to their defining plumage, they possess a rose comb and some silver-laced varieties may have gray legs. Wyandottes are known for their friendliness, as well as being broody. They are calm, cold-hardy, nonflighty birds, ideal for beginners.

Breeds by Purpose at a Glance

PURPOSE	BREED	AVERAGE WEIGHT (POUNDS/Kg)
Table	Brahma	9 lbs/4 kg
	Cochin	9 lbs/4 kg
	Rock-Cornish Cross	8 lbs/3.5 kg
	Jersey Giant	10 lbs/4.5 kg
Dual-Purpose	Plymouth Rock	7 lbs/3 kg
	Australorp	6 to 7 lbs/2.5 to 3 kg
	Orpington	7 lbs/3 kg
	New Hampshire Red	6 lbs/2.7 kg
	Rhode Island Red	6 lbs/2.7 kg
	Wyandotte	6 lbs/2.7 kg
Layers	Ancona	4 lbs/2 kg
	Andalusian	4½ to 5½/2 to 2.5 kg
	Araucana	5½ lbs/2.5 kg
	Black Sex-link	5 lbs/2.3 kg
	Fayoumi	3½ to 4½ lbs/1.5 to 2 kg
	Leghorn	4½ lbs/2 kg
	Minorca	7½ lbs/3.5 kg
	Red Sex-link	5 lbs/2 kg
Show	Crèvecœur	7 to 8 lbs/3 to 3.5 kg
	Faverolle	6½ to 7 lbs/3 to 3.2 kg
	Frizzle	4½ to 6 lbs/2 to 2.7 kg
	Houdan	6½ to 8 lbs/3 to 3.5 kg
	Lakenvelder	4½ to 6 lbs/2 to 2.7 kg
	Polish	4½ to 6 lbs/2 to 2.7 kg
	Sultan	4 to 5 lbs/2 to 2.3 kg

COLOR ME BEAUTIFUL

Chickens lay eggs in a variety of colors. Eggshells get their color from pigments that are deposited as the egg moves through a hen's oviduct. These pigments are genetically determined. Variations in eggshell color are not related to differences in egg flavor, as flavor is determined by diet, not genetics. If you are looking for a particular color, or simply wish to have an array of colors available for use, use this guide in making your selections.

Blue: Araucana

Blue-green: Ameraucana and Easter Egger

Brown (Dark and Light): Australorp, Barnevelder, Black and Red Sex-link, Brahma, Cochin, Cornish, Delaware, Dominique, Faverolle, Hamburg, Jersey Giant, Lakenvelder, Malay, Maran, Naked Neck, New Hampshire Red, Orpington, Plymouth Rock, Rhode Island Red, Welsummer, and Wyandotte

White: Ancona, Andalusian, Campine, Crèvecœur, Dorking, Houdan, Leghorn, Minorca, Polish, Silkie, Sultan, and Yokohama

FROZEN COMBS AND WILTED FEATHERS

Weather plays a significant role in the health of your chickens, as some are genetically better able to withstand summer's swelter while others will gladly frolic in the snow. Bantams generally do well in the heat of summer, except for the feather-footed varieties, whereas the heavier standard breeds will fare better in cold weather, on account of their larger size and denser plumage. The bigger the comb or wattle, the more susceptible a bird will be to frostbite. Knowing which breeds function better in varying climates can prevent discomfort on their part and heartache on yours, should a nor'easter or heat wave cause your fine feathered friend to go claw up.

Cold-Hardy Breeds
(for climates which regularly get below freezing for part or most of the year)

Ameraucana	Hamburg
Araucana	Jersey Giant
Australorp	Langshan
Chantecler	Orpington
Cochin	Plymouth Rock
Cornish	Silkie
Dominique	Sussex
Dorking	Wyandotte
Faverolle	

Heat-Tolerant Breeds

Andalusian	Leghorn
Buttercup	Minorca
Campine	Rhode Island Red
Cubalaya	Sumatra
Fayoumi	

Breeds Hardy in Both Heat and Cold

Aseel

Brahma

New Hampshire Red

RARE BIRD, INDEED!

When selecting your flock, it would be well worth considering rare and endangered breeds. The hybridized uber layers and broilers favored by industrial agriculture have displaced many heritage breeds, and many chicken varieties are now threatened or endangered. The loss of such genetic diversity does us all a huge disservice in the long run, as a reduced gene pool could spell quick and sudden extinction should an especially virulent virus swoop in.

The American Livestock Breeds Conservancy (ALBC) maintains a list of endangered breeds on their website, www.albc-usa.org. Similar organizations exist internationally, each with its own criteria for classifying endangered birds. The ALBC's conservation priority list is based on numerical guidelines as follows:

Critical: Fewer than 500 breeding birds in the United States; globally endangered.

Threatened: Fewer than 1,000 breeding birds in the United States; globally endangered.

Watch: Fewer than 5,000 breeding birds in the United States; globally endangered. Also included are breeds with genetic or numerical concerns or limited geographic distribution.

Recovering: Breeds once listed in another category having now exceeded Watch category numbers but still in need of monitoring.

Study: Breeds of interest but either lack definition or genetic or historical documentation.

Critical	Threatened	Watch	Recovering	Study
Andalusian	Ancona	Brahma	Australorp	Araucana
Aseel	Cubalaya	Cochin	Leghorn (Nonindustrial)	Egyptian Fayoumis
Buckeye	Dorking	Cornish (Nonindustrial)	Orpington	Iowa Blue
Buttercup	Lakenvelder	Dominique	Plymouth Rock (Nonindustrial)	Lamona
Campine Langshan	Langshan	Hamburg	Rhode Island Red	Manx Rumpy
Catalana	Sussex	Jersey Giant	Wyandotte	Modern Game
Chantecler		Minorca		Naked Neck
Crèvecœur		New Hampshire Red		Old English Game
Delaware		Polish		Phoenix
Faverolle		Rhode Island White		Shamo
Holland		Sebright		Sultan
Houdan				Yokohama
Java				
La Fleche				
Malay				
Nankin				
Redcap				
Russian Orloff				
Spanish				
Sumatra				

Breeds and Dispositions

Calm, Friendly, and Easygoing			Flighty, Nervous, and Aggressive	
Ameraucana	Cornish	Orpington	Ancona	Leghorn
Araucana	Dominique	Plymouth Rock	Andalusian	Old English Game
Barnevelder	Dorking	Polish	Buttercup	New Hampshire Red
Belgian Bearded d'Uccle	Faverolle	Silkie	Fayoumi	Rhode Island Red
Brahma	Java	Wyandotte	Hamburg	Sebright
Cochin	Langshan		Lakenvelder	Sumatra

Portrait of a chicken owner

Christine and her feathered friends

When Christine took up keeping chickens in the backyard of her London home, she never anticipated the indelible mark that simple act would leave on her entire life. In her previous career, and in her new status as a university student, she was always in the company of others. Faced with a stretch of quiet summer months, Christine decided to welcome three hens into her urban environment to keep her company. After purchasing a portable prefabricated housing unit in which to accommodate her lovely new friends, Christine connected with an online community for owners of the stylish polymer units. This connection in turn linked her with the charity Battery Hen Welfare Trust, which works to find new homes for factory-raised hens considered to no longer be desirable. Christine works to ensure that these hens are placed in "retirement homes" with regular families, where they are permitted to free range and live out the remainder of their lives in solace.

As a jewelry designer, Christine has found design inspiration in her small flock. Their antics and movements are always in the back of her mind when creating new pieces for her collection. For Christine, what began as a simple desire for companionship turned into so much more. As she tells it, **"I never would have dreamt that keeping pet chickens would become so influential in my life. I guess that if chickens didn't change my life, they certainly did strongly influence its direction!"**

Chapter 3
Obtaining Chickens

You've chosen your breeds, checked to see if you need a permit to keep chickens in your area, and considered the space available. Now you're ready to start assembling your motley crew! Consider whether you want chicks or more mature chickens, what you should be looking for when selecting birds, where to source your chickens, and what time of year might be best for chicken purchasing.

HERE, CHICK, CHICK!

If your local pet store is anything like most, you won't be finding White Leghorns or Japanese Bantams there any time soon. While most big box pet retailers will sell you salamanders, turtles, finches, and dog sweaters, it's pretty unlikely you'll score any Belgian Bearded d'Uccles there, either. Your local feed store is a much more likely place to begin feathering your nest. Selection there may be limited, however, especially if you are after a specific exotic or endangered breed. Feed stores can offer a wealth of chicken advice, though. Ask a staff member who supplies them with their poultry and whether their suppliers are nationally recognized as engaging in sound and healthy breeding practices. Participants of such organizations often insure their flocks to be free of once-common diseases such as pullorum and typhoid, nasty bugs your chickens can literally live without.

Mail order catalogs are a great source for obtaining chicks.

If you don't have a feed store where you live (in, say, midtown Manhattan), or you have your heart set on a Lakenvelder that can't be had at John & Jane Doe's Feed & Seed, then move your search to the Internet. It is possible today to source chickens and chicks online from a number of highly reputable hatcheries. Some offer free catalogs, should you want to take your time ogling and contemplating chicken varieties over your morning cup of joe. These mail-order hatcheries also often sell housing, feed, and reading materials for those desiring a one-stop chicken shop.

Another place to look is the classifieds section of your local newspaper. City newspapers sometimes offer treasure troves of local poultry for sale. Additionally, you may want to seek out poultry clubs in your area, which can be found through a quick Internet search. Club members can recommend trusted chicken suppliers, either in your area or online. A local veterinarian can also be a good source of information and referrals.

Finally, don't underestimate the county fair as a means of procuring a desired breed. Owners of prizewinning birds, or any of the birds on display, might be in the business of selling offspring or could be talked into doing so. Take a gander, see who the chickens belong to, and try to track them down, either by the names listed on the bird's cages or through the fair's livestock coordinators. Score some chickens and ride the Ferris wheel—now that's my idea of a good time!

AGE APPROPRIATE

When you begin assembling your flock, you will have to decide how old your chickens will be when you bring them home. It is possible to purchase freshly hatched chicks, pullets (usually between 16 and 20 weeks old), or mature hens. If you are considering including a rooster in the mix, they may also be purchased as chicks, cockerels (male chickens less than one year old), or mature roosters. Although it's hard to resist the ridiculously cute, fuzzy balls of feathers that are chicks, this decision should not be made lightly, as there are pros and cons to each scenario.

Chicks

If you want the most birds for your money, chicks are the way to go. Depending on the quantity you purchase, chicks usually only cost a few dollars apiece. Some hatcheries require bulk orders of chicks in order to ensure they will remain warm in transit. However, as backyard chicken keeping is on the rise, hatcheries are becoming more responsive to the needs of the small grower. Some will even offer single chicks for purchase.

Chicks bought from a reputable source, either your local feed store or an online commercial hatchery, will generally be in good health when you get them. If you intend to keep your chickens as pets, and not merely as table birds, acquiring them as chicks enables you to develop a relationship with them, building trust as you hold them and feed them by hand. Should you be so lucky, you might one day end up with a chicken that will sit in your lap or perch on your shoulder!

Before you rush to your nearest feed store and gather up the fluffiest chicks you can find, remember that, at least in the beginning, chicks do require more care than older birds. During their first few weeks of life, chicks are at their most vulnerable. Like any young animal, chicks can be rather unruly. Left to their own devices, things could get pretty messy, and potentially hazardous, fairly quickly. They frequently stomp around in their food and water, spreading fecal matter throughout. Savagely enough, they can peck each other to death. Be certain before you purchase your chicks that you will always have someone available for "chick patrol." Furthermore, they are susceptible to a host of illnesses and health conditions.

We will discuss newly hatched chicks and their particular needs in greater detail in chapter 7. Raising chicks requires a significant time commitment. If you don't think you will be able to provide them with what they need, you may want to consider purchasing pullets instead. Conversely, if you do have

time to be a chick nanny, little can compare with the hilarity of watching chick antics and the joy of hearing their little chick peeps.

Another thing to consider is that, unless you purchase sexed chicks, where the sex is determined for you by a poultry sexing expert, the precious chicks you brought home might leave you with several roosters too many, which you don't want. If chicks are your choice, be certain to request sexed chicks and develop an appropriate rooster-to-hen ratio. (Unsexed chicks are also listed as "straight run" and "as-hatched.") If you live in an urban or suburban area, remember that roosters are most likely prohibited, so choose your chicks accordingly.

Started Birds

Although they will cost more than chicks, usually by several dollars apiece, pullets and cockerels require a good bit less care than newly hatched chicks. If your goal is to have a regular supply of fresh eggs to eat, pullets will be that much closer to point of lay—by six months they may already be laying. Accordingly, they will cost you less in feed as they mature. Birds purchased as starts will offer more eggs over their lifetime than those purchased as hens. Also, started birds require less in the way of specific housing and care than do chicks. Pullets can immediately be placed in their run, needing no regulated heat or special chick feed.

On the other hand, started birds are often not as readily available at feed stores as are chicks. If you plan to raise chickens for table, started birds are too old, as table birds are generally slaughtered around eight to 12 weeks. Another down side is that, not having been hand raised from birth, pullets may not be as tame as they might have been had they grown accustomed to you and your environment over time. Furthermore, you may not be fully aware of how an older chicken was handled prior to moving in with you. If you want to ensure yourself stewardship of affable, sociable, guest-and-family-appropriate chickens, then selecting chicks from a breed known for their friendly personalities may be your best bet.

Gender Studies

Professional chicken sexing experts determine chick sex. This occurs by careful observation of the vent for minute gender differences. (The vent is the eliminatory organ through which eggs and feces pass, as well as a chick's genitalia.) These differences are remarkably slight, and chicken sexers develop accuracy through repetition in observation. Even then, sexers are right about 90 percent of the time. This method of gender differentiation is used mostly by large hatcheries. In my case, of the five chickens I received from my nurse friend, all five grew up to become hens. Unfortunately for her, of her five chicks, two turned out to be roosters!

Mature Birds

When electing to purchase a mature bird, as either a hen or a rooster, the clear advantage is that you generally are able to see precisely what you are getting. Older birds are clearly sexually mature and have their plumage in. Like pullets, they require less in terms of housing, critical care, and constant observation than do chicks.

Mature hens, although they will continue to lay, will never produce as many eggs after their first laying season. Older birds are also more susceptible to diseases as they age. Finally, purchasing mature birds will cost considerably more than purchasing chicks, as their entire lifetime up to that point will have been maintained on someone else's dime.

SHINY EYES AND SLICK FEATHERS

When purchasing chickens, be on the lookout for visible cues to the birds' health. If you are purchasing chicks by mail, be certain to open the box in front of a delivery person to make sure your flock all survived the journey. I would not advise bringing children along with you on this errand, as a limp, deceased chick is not an image you'd like to have etched into your child's memories. If purchasing chicks in person, look for alert, energetic birds. Pick them up and examine their rear-ends for pasty butt, which is exactly what it sounds like, a backside with excrement dried around it, preventing elimination. Be certain to choose chicks with straight beaks and toes.

Older chickens should have clear, bright eyes, waxy combs and wattles, shiny feathers, and smooth legs. There should be no visible parasites, which an examination under the wings and around the vent should easily disclose. Internal parasites will cause diarrhea, which a quick check at the vent should indicate. Pick up any bird you are thinking of buying. The breast-bone should be flexible and covered with flesh. Listen for any coughing or wheezing, as this could be an indication of a sick bird. Examine the entire flock, looking for any listless or isolated birds. One sick bird could affect the entire crew, so pay close attention. Lastly, if you can, try to have a look at the bird's droppings. Healthy birds will have firm, well-rounded feces, brown in color and tipped in white (this is the urine).

WINTER, SPRING, SUMMER, OR FALL?

It is important to consider the time of year when purchasing your chickens. The season in which you assemble your flock will affect their behavior, needs, and laying habits. If purchasing chicks by mail, they will need to be kept warm in transit from the hatchery in cool months. They will then need to remain warm on the way home from the post office or feed store. Feed stores generally

have chicks available for purchase in the spring, while most mail-order hatcheries offer chicks every season except winter.

Pullets and mature birds will need to be kept dry in cold, wet weather, and their egg laying will drop off as the daylight hours shorten. In very cold climates, some chickens' combs will need to have a lubricant rubbed on them to prevent frostbite. Bantams especially will need to be kept out of heavy snow and wind during cold months.

During warmer months, chicks and older birds will need to have access to shade and water. They may also go broody during this time of year, which you may need to discourage if you don't want chicks. Brooding chicks is best done in early spring, as the weather will become increasingly warmer but will remain cool enough in the evenings to ward off diseases. Late summer to early fall is when most chickens molt. Egg laying will lessen or cease altogether during this period.

The Chick's in the Mail

For the best selection, you may opt to purchase chicks by mail. While delivering a box full of live chicks may be an out of the ordinary occurrence for an urban postal worker, rural post offices are accustomed to express shipments of peeping feather balls. Chicks are shipped by air, with 48 hours elapsing from hatching to delivery at your post office. They are sent in a specially designed chick-shipping box equipped with ventilation holes. It is rare for most hatcheries to ship fewer than 25 chicks per order. The high number is needed for chicks to huddle together and generate body heat, necessary to their survival. If this is the policy of the hatchery you choose, you may want to seek out other members of your community interested in purchasing at the same time. Hatcheries do exist that will send smaller numbers of chicks, but they may either charge an additional insurance fee or throw in extra male chicks for added warmth. You may want to give your local post office a call with a heads-up about your imminent arrival. They will call you once your package arrives either way.

Portrait of a chicken owner

Robin

City councilwoman, musician, chicken keeper—it's all in a day's work for Robin. She's just as content rubbing elbows with the mayor and citizens of her small city as she is throwing out scratch and stroking silky feathers. Her flock of six hens was acquired through a bulk order with friends in order to save on both shipping and individual bird costs. She loves her ladies, but they haven't turned out to be quite the layers she'd hoped for. Next go 'round, she'll be sure to go for those breeds known for their laying abilities. Asked for a nugget of urban chicken-keeping wisdom, Robin reminds all to **"feed 'em regularly and early so they don't wake the neighborhood."**

Chapter 4
Housing

Although chickens don't need much square footage (they are rather diminutive birds, after all), they do require appropriate housing. Setting up accommodations will be the largest part of your start-up costs. Once established, the housing should last for a number of years, if cared for properly. The essentials of any henhouse include perches, nesting boxes, security from predators, and access to the outdoors. Beyond the basics, you can customize your flock's housing in an infinite number of ways. Chicken palace or humble abode, the choice is up to you. A number of variables must be considered, though, before hitting the first nail or rolling out the first length of chicken wire. Size of flock, climate, and available space are just a few of the factors that must be taken into account when setting up your coop de Ville.

FOR HERE OR TO GO?

First determine whether you want permanent or mobile housing for your flock. There are advantages and disadvantages to each choice. Permanent housing will be worth considering if you plan on being where you are for some time. If you live in an area subject to extreme heat or cold, permanent housing will provide the greatest amount of protection from the elements. It can be wired for electricity to power heat lamps, water warmers, and ventilation fans, in addition to overhead lighting. With a bit more fortification, permanent housing will also be more predator proof than many mobile setups. Additionally, a greater number of birds may be kept in permanent housing.

Permanent housing can be considerably more expensive to build than mobile housing. If you are planning on building the coop yourself, you may be able to reduce some of the costs. However, providing good insulation, ventilation, fencing, and predator-proof protection will add to the expense, whether you are the builder or not.

Mobile housing may be the ideal choice for those with limited space or budgets. If you intend to keep a small flock, in both number and physical size, and your climate is mild, this may be the way to go. Mobile housing also can help with yard maintenance. Lightweight and bottomless, chicken "tractors" or "arks" are structured to be moved to any location needing weed control and soil enrichment. Your flock gets access to grass, insects, worms, and other slithery soil inhabitants chickens love to dine on, and your yard will be all the greener. If you have flighty birds such as bantams, you won't have to worry about installing pricey tall fencing, as mobile housing is completely enclosed. Furthermore, small mobile coops are often easier to keep clean than larger permanent housing.

If you live in an urban or suburban area and wish to have the ability to move your chickens periodically to avoid having them too near any particular neighbor, mobile housing can be ideal. You may construct your own ark relatively inexpensively, or you may purchase a manufactured version from an online seller. A growing number of mobile pre-fabricated housing units are available in a variety of colors and modern designs. As backyard chicken-raising fever catches on, the design world is

A mobile pre-fab chicken enclosure

rising to the occasion, meeting the discriminating preferences of today's chicken tenders.

BREATHING ROOM

Proper ventilation is essential to chicken health. Chickens have a high rate of respiration and subsequent production of carbon dioxide and moisture, which makes them particularly susceptible to respiratory diseases. Keeping air moving through their coop is therefore fundamental to their vitality. Inevitably, moisture will be introduced into the coop from rain, as well as the chickens' own breath and droppings, and you will need to take steps to remove it.

Proper ventilation is essential in the coop.

If you are designing a permanent structure, make sure that at least one-fifth of the wall space has windows that open for air circulation. If you opt for glass windows, simply raise them. Cover window openings with ½- to ¾-inch (1.3 to 1.9 cm) galvanized mesh wire on the interior of the coop to prevent interlopers from accessing the coop. Predators can be extraordinarily tenacious in their drive to get into chicken coops; further methods of fortification will be discussed below. Wild birds, which could carry parasites or diseases, must be kept out, as well. Use the right wire in the beginning and save yourself, and your feathered friends, from unnecessary suffering in the long run.

In a small or mobile structure, windows may not be an option. Instead, situate ventilation holes on the north- and south-facing walls near the top of the coop. Cover the holes on the inside with mesh wire. Install a wooden drop-down cover over the screens with hinges at the bottom and a latch or hook at the top. This way you can monitor and adjust the ventilation inside the coop as needed, while keeping wildlife from gaining access. Only close both covers if the weather turns terribly cold. Otherwise, leave the south-facing airholes open at all times, even in chilly weather.

In warmer weather you will need to ensure constant cross-ventilation for your birds, especially for those that spend most of their time in their coop and have minimal outdoor access. This can be achieved either by opening all doors, windows, and covers, or through the use of a fan suspended from the ceiling of the coop or mounted to a wall. Chickens aren't capable of sweating and can go downhill pretty quickly in temperatures over 95°F (35°C). You'll know they are too hot if you see them panting. Keep their coop well ventilated and give them access to the outdoors, shady spots to hide from searing sun, and fresh water at all times, and your flock should cluck along happily all summer long.

TIP:

The Nose Knows

To determine whether your coop is properly ventilated, trust your nose. When you enter the coop, take a big whiff. Do you smell ammonia? If so, your coop is not properly ventilated.

SIZING THINGS UP

While bigger isn't always better, having more coop space to start is certainly better than having too little. The less crowded your flock is, the healthier and more content they (and you) will be. You may want to increase the size of your flock over time, and having planned for expansion at the start will save headaches in the future.

That said, there are no hard and fast rules about coop size. In my research, I've found guidelines suggesting everything from 4 to 10 square feet (1 to 3 m²) per bird. Variations account for the size of individual breeds, the number of birds you are keeping, and whether your birds are given access to a large outdoor area or are kept confined in small quarters most of the time. The more space you have available to you, the bigger you can make your coop, if you are so inclined. In my case, I live on 12 acres (48,000 m²) in a relatively rural area. The size limitations of our coop were restricted only by our imaginations! We appropriated an existing 64-square-foot (6 m²) dog shelter built on a platform, put walls on it, installed roosts and nesting boxes, and created essentially a chicken palace for our five Ladies. They also have access to an

800-square-foot (74.3 m²) fenced run, another vestige of the dog shelter. That is an inordinately large amount of coop and run space for a flock so small. It was what we had available, though, so we simply improvised and piggybacked on an existing structure.

Most folks living in urban, and perhaps even suburban, areas will have limitations on the number of birds they can keep. It would be worth asking town officials about coop size restrictions as well. Before you make your first trip to the hardware store, determine if a building permit will be necessary. More than likely, if you keep your coop small, you won't need one.

For birds with outdoor access, allow 2 square feet (.19 m²) per bird in the coop and 4 square feet (.37 m²) per bird in the run. This will provide ample room for your flock, as they will spend most of the daylight hours outside the coop, hunting, scratching, chatting, fluffing, and doing all the glorious things chickens do. Those numbers can be halved for bantams. For birds kept confined most of the time, or perhaps during lengthy periods of inclement weather, double the interior space, up to 4 square feet (.37 m²) per bird. You will also need to allow for more space if you are keeping heavy breeds, as they simply take up more room.

If at all possible, intensively confined birds should have access to grass. One option is to keep the birds in a tractor and move around each day, returning them to their sleeping quarters at night. The flock's health will fare much better if they are given space to roam and hunt for insects. If this is not an option, however, just be certain to give them a daily ration of leafy greens (lettuce, kale, chard—my Ladies love parsley) in addition to their feed.

In addition to allowing for the birds' needs, be certain to consider your own. Pity the poor chicken tender who has to stoop low to gather eggs and scrape poop! If you are making a larger permanent structure, plan for a person-sized entry door. Otherwise, make sure your small coop or mobile unit has easy access to nest boxes and floors. Plan wisely and you won't need to be a hunched-back chicken keeper.

Some unseemly chicken behavior may result if proper space requirements are not provided. Pecking, biting, egg-eating, and sometimes even cannibalism can be the lot of a flock kept in too close of quarters. Humans get this way too, usually minus the cannibalism, if packed in together too tightly. Empathize with your crew, and give them the space they need.

KEEP YOUR FRIENDS CLOSE AND YOUR ENEMIES OUT

Predator-proofing your henhouse is of utmost importance. I've heard tales of sly foxes who accessed coops, killing 30 chickens and only taking one. There is little more you can do in terms of flock protection than fortifying their sleeping quarters against four-legged threats. Clever predators can undo latches, find the one crack you failed to notice, and shimmy down roofs. Be mindful of areas of structural vulnerability and you save your flock, not from only stress, but potentially from an untimely demise.

A well-protected chicken coop

Depending on where your flock is housed, danger can come from above, below, or both. Free-ranging flocks are subject to flying predators such as hawks, owls, and eagles, while confined flocks face burrowing predators such as weasels, raccoons, foxes, possums, snakes, and rodents. Most predators stalk for prey at night, or in some cases, dusk or dawn. If you allow your birds to pasture, watch out for hawks, the most brazen daytime predator. Camouflage a pasture flock by purchasing darker colored breeds and ensuring there are bushes or other hiding places for them to run to when trouble soars overhead.

To keep the henhouse safe, think like a predator and take action before they can literally "weasel" their way in. Close up coop doors and windows after the chickens have roosted for the night. Reinforce latches and gates with strong, complex locks. It might be worth placing two types of latches on gates and doors, one at the top and an additional one at the bottom. If a toddler can easily open your closure, so can a predator. Some predators are simply looking for feed, not for chicken meat. If you find your feed has been accessed, either secure it tightly with bungee cords, build a wooden box with a secure lock to keep it in, or store it indoors, scooping out what you need for the flock on your way out each morning.

Deter predators by keeping the grass surrounding your run mowed, and remove nearby piles of debris or objects. Roofs will help keep flocks safe from climbing and flying predators, as well as keep in flighty birds. Permanent housing should have ½- to ¾-inch (1.3 to 1.9 cm) galvanized wire mesh covering all ventilation holes; 1-inch (2.5 cm) chicken wire is too large and will allow a number of predators access. Raccoons can easily pull out regular staple-gun staples, so be sure to use wood staples affixed with a hammer when constructing your housing. Mobile housing should be moved daily and have wire surrounding it that is small enough to prevent weasels and mink from climbing deftly through.

Lastly, if your state permits it, you may want to consider trapping and relocating sneaky pests. You will need to first determine what is getting into your coop and then trap accordingly. If you plan on trapping the predator yourself, please do so in as humane a manner as possible. Remember, what we consider "predators" are simply living creatures looking for food, or in the case of some dogs, playing too aggressively. Check traps several times a day. If you discover an animal, cover the trap with cloth in order to calm it. Either transport the animal to your local animal control facility or relocate it at least 5 miles from your home. If you trap more than one of the same species, relocate it to the same area. If you are attempting to relocate and release a wild animal by yourself, read all trapping instructions closely before beginning, and be sure to wear gloves. You may be able to borrow a trap from your animal control and wildlife agency instead of purchasing one yourself, so it might be worth it to call and determine whether this is an option in your area.

FENCE ME IN

Fencing can be an invaluable tool in protecting your flock. If installed properly, it can go a long way toward keeping your crew in and sneaky beasts—intent on ravaging chickens, their offspring, and their eggs—out. Fencing should be no less than 4 to 6 feet (1.2 to 1.8 m) high, depending on the flightiness of your breed.

Honeycomb-pattern chicken wire has its place, but not really as fencing. It is flimsy and acts as little deterrent to a predator determined to get in. It works better as overhead cover from predatory and wild birds in a run. Galvanized chicken wire intended for outdoor use may be suitable for use as fencing. Check how sturdy it is when considering it for purchase.

Proper fencing keeps chickens in and predators out.

A typical latch-type gate closure

YOU LIGHT UP THEIR LIVES

Chickens need light. The break of dawn is their cue to lay eggs, as morning's light sends a signal to a hen's pituitary gland, which in turn informs her ovaries to get busy. The darkening at day's end tells the birds to return to the coop for some shut-eye. As their caretaker, you will need light when tending to your chickens. A well-lit coop lets you clean properly, check for eggs, and count your flock for any missing.

During winter, when daylight dips to less than 14 hours per day, chickens may reduce their laying or cease to lay altogether. Having windows in their coop and access to sunlight outdoors may temper this natural response, although it may not prevent it entirely. If sporadic laying simply translates to a few less omelets and custards for you and is no big deal, then it's probably best to let nature take its course. If you are committed to having a specific number of eggs year-round, however, you will need to install artificial lighting.

One 40-watt bulb placed 7 feet (2.1 m) off the ground is plenty of light for a 100-square-foot (9.3 m²) coop; opt for 60 watts if your coop is big-

Adequate light is essential.

ger. In order to mimic longer hours of daylight, you will need to turn the lights on either in the morning or the evening to achieve a total of 15 daylight hours. I recommend extending your chickens' day at dawn as opposed to dusk. If you turn out the lights on your flock too quickly, before they're settled onto their roosts, you'll scare the…well…poop out of them unnecessarily. Either trust yourself to head out to the coop earlier each day or set up a lighting timer to do the work for you.

Ideally, fencing should be made of heavy-duty yard or livestock fencing or netting. Holes in the wire should be no larger than 1 inch (2.5 cm) so that chickens cannot stick their heads out and chicks cannot easily slip through them. You may want to consider "double fencing," which consists of an outer layer of heavy-duty fencing and an interior layer, perhaps only a few feet high, of more traditional chicken wire. This helps keep predators out and your feathered crew in. It is advisable when erecting the fencing to bury it 6 to 12 inches (15.2 to 30.5 cm) below ground, bending the end to a 90° angle. By the time a burrowing predator or digging dog reaches the angle, they will most likely determine it's a lost cause and give up. Attach the fencing to posts set about 4 to 6 feet (1.2 to 1.8 m) apart. Posts will need to be buried about 6 to 8 inches (15.2 to 20.3 cm) into the ground, so take that into account when determining the length of posts you need.

Alternatively, consider electric fencing. Affix insulators 6 inches (15.2 cm) from both the top and the bottom of your wire fencing, then thread electric wire through the insulators and around the fencing perimeter. Top wire placement thwarts predators thinking of scaling the fence, while bottom wire deters those thinking of tunneling in or lurking around the exterior. Transportable electric fencing is an option for free-range flocks ("mobile poultry fencing" is the term to search for online or in catalogs). Made of plastic string laced with fine metal wire, netting comes with posts built into it for sturdiness. Mobile fencing may be battery powered or plugged into an electric outlet.

Pullets need fewer hours than layers, only 8 to 10 as opposed to 15. Giving pullets too much light too soon may cause them to mature too quickly, which can result in prolapse. Prolapse is a condition where tissue just inside the vent is forced out. Eggs may remain attached inside. Remove a prolapsed pullet from the flock as soon as possible, as the visible tissue may encourage other chickens to peck, causing her to hemorrhage. Apply a lubricating hemorrhoid cream, push the tissue back inside the vent, and isolate the pullet until she is healed. Or, save yourself the trouble and use artificial lighting only with layers, never with pullets.

WELL-APPOINTED BEDDING

Bedding, also known as litter, is the absorbent material put down in nesting boxes and poultry housing floors. It is used to provide a cushion for your bird's feet and eggs, offer them insulation, minimize their contact with fecal matter, and absorb their droppings. Any number of materials would be suitable as bedding.

The most important thing to consider when selecting appropriate litter is that it is completely dry. Any trapped moisture

could become a breeding ground for mold and bacteria, which can in turn affect your flock's respiration. Each day, rake over the bedding and, over time, it will break down and begin to compost. You will need to remove the bedding completely several times a year in order to prevent the spread of disease or in-

fection, as well as provide a clean and happy home for your flock. Even chickens appreciate a thorough spring (and fall) cleaning. Appropriate bedding materials for feathering your girls' nest include wood shavings, sawdust, chopped straw, ripped-up newspaper, dry leaves, even peanut shells. If you opt for wood shavings, be certain to use those designed specifically for this purpose, as some shavings contain fine dust particles that may cause respiratory problems in chickens. Whatever you choose, be sure to put it down at least 2 inches (5 cm) thick in the nesting boxes and 5 to 10 inches (12.7 to 25.4 cm) deep on the floor. For my Ladies, I've put down cedar shavings as bedding. Each nesting box contains shavings about 3 inches (7.6 cm) deep, while the floor has about 6 inches (15.2 cm). Every morning, after I let the Ladies out of the house and feed them, I grab a garden rake kept hanging from the henhouse wall at all times and give the bedding a quick stir. It covers up the droppings and evens out the smell. In fact, my henhouse almost always smells great!

The method I use, often referred to as the "deep litter" bedding method, allows for the convenience of needing to change out bedding completely only every four to six months. You will need to replenish supplies along the way to maintain a layer between 5 and 10 inches (12.7 and 25.4 cm) at all times. The thickness of the bedding layer, well aerated, will help prevent odor. Some chicken keepers even throw grain on the housing floor to encourage their chickens to scratch. Given incentive, the chickens themselves will stir up the droppings and work them into the bedding. I mean, really, why should you do all the work when you have beaks and claws available to assist?

Alternatively, some chicken keepers use what is referred to as a droppings board or pan. A mesh-wire covered pan or wood board with raised slats is placed underneath the roosts. This allows droppings to fall down through the openings in the board to a place where chickens cannot peck at them. If you use this method, be sure the droppings board is high enough off the ground to be inaccessible to burrowing animals. Whatever method you elect, remember that chicken poop makes great fertilizer. When you remove droppings and litter, simply toss them into a compost pile. From eggs to meat to garden amendments, chickens are the pets that just keep on giving!

A FOUNDATION TO BUILD UPON

Proper flooring is essential to a safe, healthy, and solid coop. If your housing is mobile, then your "flooring" is simply grass. As long as you rotate the location of your tractor every day, grass flooring will work well. Commonly used types of flooring for permanent housing include concrete, wood, dirt, and the aforementioned droppings boards.

Concrete flooring offers the greatest degree of protection from rodents and burrowing predators, and it's easy to clean. You may elect to mix and pour the concrete yourself or hire a contractor to do the job for you. Either way, concrete is your most expensive flooring option. Concrete isn't a material you can simply move should you become dissatisfied with your chosen location, so make sure your site selection is final before you pour.

Wood floors are great for thrifty builders; you can really use just about any wood to do the job. Recycle wood from an old project, scavenge some from someone's curbside rubbish, or simply buy new from the local hardware store. Your floor will need to be raised off the ground by at least 1 foot (30.5 cm), on either posts or concrete blocks. The downside to wood floors is that they can be difficult to clean and, accordingly, will need to be replaced every few years, based on regular wear and tear.

One solution that I implemented in my henhouse is to line the wood floor with linoleum. Our friend Tom, an amazing handyman if there ever was one, swooped up the linoleum in the corners when he first helped set up the coop, allowing the linoleum to rise up about 5 to 6 inches (12.7 to 15.2 cm) above the floor. This way, come cleaning time, all I have to do is sweep out the deep litter into a wheelbarrow (bound for the compost heap), hose down and mop the floor, and then spread a new layer of deep bedding. No nasty scrubbing of wooden floorboards!

You may also place your coop over a simple bare earthen floor. While this is seemingly the most cost effective option, it may cause more trouble, and subsequent expense in repair, in the long term. Dirt that is not sandy will simply become too moist and muddy during wet weather. Remember, excessive moisture is public enemy number one with chickens. While a dirt floor will help keep chickens cooler during warmer months, it will pull heat away during cooler times of the year. Dirt floors can also be a challenge to keep free of droppings and can by no means be considered predator or rodent proof. If the dirt where you intend to situate your housing drains well and is sandy, however, such flooring may be a good choice for your needs. You will need to rake the floor daily to remove droppings and cover over holes your chickens may have dug. You may also need to place wood chips or sand atop the dirt to prevent it from getting too mucky.

Chicken housing comes in all shapes, sizes, and materials.

SITE SPECIFIC

Choosing the proper location for your housing is impera-tive. If you pick a bad spot, you could be faced with multiple inconveniences to both you and your flock. Ideally, permanent housing should be situated on a south-facing slope or slight incline to allow water to drain away during rain and snow, as well as permit ample light for drying out the run and warming up the coop. Windows placed on the south-facing wall let light flood the coop, so your chickens receive sunlight even on short winter days.

Make sure that your coop is located close enough to your home. You will want to be aware of the goings-on in the coop, keeping a watchful eye out for predators and other dangers, including those the chickens might bring on themselves. Also, having your flock nearby allows you to bring them water and food without having to travel great distances, which on cold winter mornings or during wet spring weather will be a god-send. If your coop is not wired, having the housing adjacent to your own will allow you to rig up electric cords, should you have need for them.

Additionally, you will want to make sure that your flock's housing is situated both according to ordinances and setback codes where you live and not too near to your neighbors' homes. Even the tidiest of chicken tenders cannot escape the "aroma" generated by daily chicken droppings. Making certain the coop's site is far enough away from your neighbor's windows will go a long way towards engendering goodwill. I once read of a clever, and pre-scient, chicken keeper who strategically planted sweet-smelling perennial plants around her coop, allowing the heady perfume of the plants to override the malodorous scent of chicken funk.

If you allow your chickens to free-roam, be mindful of the existing landscaping in your yard. Chickens are not known to traipse gently through the great outdoors, preferring instead to scratch, peck, and otherwise trample whatever they come across. If you have tender shoots or seedlings, prizewinning roses, or carefully manicured borders you would prefer remained as such, plan to either limit the chickens' yard access with a run, or purchase transportable fencing and situate them precisely where you want them to be.

KEEP IT CLEAN

Regular cleaning practices are essential to maintain the structural integrity of your housing and keep parasites at bay. While minor maintenance will need to happen daily or monthly, the "big clean" need only occur once a year. The number of birds kept dictates how often you will need to clean; small flocks are easier to keep tidy than large, hundred-bird chicken enterprises.

Keep tools at hand for keeping your coop clean.

See the appendix (page 128) for a checklist of cleaning and maintenance tasks to be done daily, weekly, monthly, biannually, and annually.

If you have droppings boards, they need to be emptied daily. Failure to do so can result in both a smelly house and hard, caked-on droppings when you finally do get around to emptying them. Those using the deep bedding method will need to rake the droppings over each day, spreading the bedding across the entire floor. Add more bedding over time to maintain a height of around 5 to 10 inches (12.7 to 25.4 cm). Empty out and replenish all the bedding every four to six months. If you use straw in your henhouse, it will need to be removed weekly, especially those areas that become moist and matted. Also, check nesting boxes to see if they have become soiled or need bedding added to them.

Check feeders and waterers daily for droppings or caked up mud. I keep a scouring pad in my coop and give the rim on the Ladies' galvanized waterer a quick scrub-down every few days to remove accumulated dust and debris. A daily scrubbing of roosts and perches is a good idea as well. I use a rake to give a quick run over the perches each morning to prevent the build-up of excrement and allow droppings to begin to compost in the bedding below. A daily perimeter check will keep you abreast of potential annoyances before they develop into problems. Look for indications of tunneling, tears in the fencing, protruding pieces of wood or nails, and any other needed repairs.

Gear up for the "big clean" once a year. It is a good idea to do this in the spring, before the heat of summer brings out every imaginable parasite. Place your flock safely outdoors, preferably on a sunny day, which will allow for rapid drying in the henhouse. Wear a breathing mask, as a lot of dust can get stirred up during the cleaning process. After you have swept out all bedding and debris, make a solution of one part bleach, one part liquid dish soap, and 10 parts water. I would advise using a non-chlorine bleach along with a natural, nontoxic, biodegradable dish liquid for this job. Toxic chemicals used for cleaning can affect your flock's health, and runoff can pollute local groundwater supplies.

Rent or buy a pressure washer or a backpack sprayer from a hardware retailer, or purchase a small unit for a nominal expense. Fill it up with the solution and saturate every surface of the interior of your coop, taking care to look for nooks and crannies that could easily go unnoticed. Leave all doors and windows opens to dry. Be certain to prevent your flock from entering during the cleaning and drying period.

BUY OR D-I-Y?

These days, it is possible to purchase pretty much everything you need to house your flock. Roosts, nesting boxes, chicken tractors, even complete coops can be purchased through mail-order catalogs and online. If you have the time and tools, you can also craft these necessities yourself.

BUILDING ROOSTS

Chickens prefer sleeping on perches, or roosts, in the evening. After all, they are descended from wild birds, which sleep in trees. While most chickens can't quite make it up into trees (not for lack of trying!), they do feel safer elevated off the ground. Wood is the best material for constructing roosts. Plastic or metal surfaces will not work, as they lack the texture needed in order for chickens to grip the perch with their toes. An old wooden stepladder or tiered towel drying rack will more than suffice. If you opt for new wood, purchase rounded dowels or round off the corners of a square piece with a sander. Roosts should be between 1 and 3 inches (2.5 and 7.6 cm) in diameter, depending on the size of your birds.

If you need more than one perch, position the dowels or pieces of wood in a stair-step fashion with levels at least 12 to 18 inches (30.5 to 45.7 cm) apart. Chickens poop at night and their droppings would otherwise fall right on top of their roommates below. Position roosts 2 to 4 feet (61 to 121.9 cm) off the ground and 18 inches (45.7 cm) from the closest parallel wall. Make sure to secure them firmly to prevent them from turning or falling down.

How Much Space?

Roost Width	Roost Length
1 inch (2.5 cm) for bantams 2 inches (5 cm) for standards	8 inches (20.3 cm) per bird for lightweight breeds and bantams 10 inches (25.4 cm) per bird for heavy breeds

Homegrown Fertilizer

Chicken manure offers a rich, natural, and free source of soil enrichment. Gardeners pay top dollar for bags of composted manure purchased from the home and garden store. With your own flock of chickens, you now have access to premium fertilizer all year round for no additional cost. Fresh chicken manure is high in nitrogen, and direct application to many plants can "burn" them, although several, such as berry bushes, welcome the heat. For most garden and landscaping plants, a better approach is to toss your chickens' droppings, along with any bedding you are changing out, into your compost bin. Through the alchemy of composting, all that nitrogen gets transformed into a less scorching, abundantly nourishing, organic fertilizer. Composting also helps kill off any harmful bacteria and viruses that may have been lurking in fresh droppings. Alternatively, you can place the fresh droppings and bedding on top of vegetable or flower beds prepped for winter rest. You could even allow your flock to romp around in the area, letting their scratching, digging, and pecking work the manure down into the soil.

eft: **Tree branch roost** *middle:* **Painted dowel roost** *right:* **Step ladder roost**

One nesting box for every four hens ensures adequate laying room.

BUILDING NESTING BOXES

Building your own chicken nesting boxes is not much harder than building a simple box. There are infinite variations on how to do it, but this is a simple version that works great. Most poultry experts suggest one box per four laying hens. The accompanying photo is of the nesting box I have in my coop. We made it larger than necessary so that we can expand our flock in the future. The project below is for a three-hole nesting box, which will easily accommodate 12 chickens.

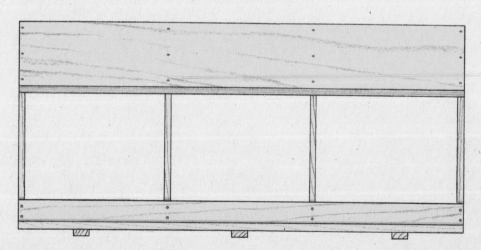
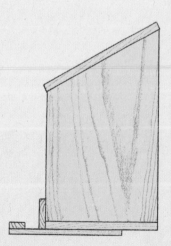

CUT LIST & SUPPLIES

Solid Wood

Number of Parts	Dimensions of each piece	Description	Comments	Ref.
2	1 x 12 board, 4' (1.2 m) long	Top and bottom of nesting box	See the **IMPORTANT NOTE** below on lumber dimensions.	A
4	1 x 12 board, one end cut straight and one cut diagonally to make one side 12" (30.5 cm) long and one 18" (45.7 cm) long	Dividing panels	You'll need about 6' (1.8 m) of 1 x 12 wood in all to make these pieces.	B
1	1 x 3 board, 4' (1.2 m) long	Nesting box lip		C
1	1 x 2 board, 4' (1.2 m) long	Perch railing		D
3	1 x 2 board, 12" (30.5 cm) long	Perch supports		E

1/4-inch (6 mm) Plywood

Number of Parts	Dimensions of each piece	Description	Comments	Ref.
1	4' x 18" sheet (1.2 m x 45.7 cm)	Optional backing		F

Other Materials and Supplies

Number of Parts	Dimensions of each piece	Description	Comments	Ref.
4	2" x 2" (5 x 5 cm)	Angle braces		
	1 3/4" (4.4 cm), 1 1/4" (3.2 cm), and 3/4" (1.9 cm) long	Wood screws		

IMPORTANT NOTE: The dimensions for boards given here are the *nominal* dimensions they are sold by in lumberyards, not the actual dimensions. The actual dimensions are as follows: 1 x 12 = 3/4 x 11 1/4 inches (1.9 x 28.6 cm); 1 x 3 = 3/4 x 2 1/2 inches (1.9 x 6.4 cm); 1 x 2 = 3/4 x 1 1/2 inches (1.9 x 3.8 cm).

Framing the Boxes

1. Place the square end of the four angle-cut boards (B) on one of the 4-foot (1.2 m) 1 x 12 pieces (A). This will be the bottom of the nesting box. Two of the boards will go on top of the 4-foot (1.2 m) piece, flush with the edge at both ends, and the other two boards will be placed about one foot (30.5 cm) from each end. Draw pencil lines on the 4-foot (1.2 m) 1 x 12 to show where the angle-cut boards will be, then take the four boards off the bottom piece.

2. Attach the angled boards one at a time, each with three 1 3/4-inch (4.4 cm) wood screws through the bottom of the 4-foot (1.2 m) 1 x 12, two near each end and one in the middle. It is best to drill pilot holes first. Clamp the board to a worktable or have a partner carefully hold the boards in place as you work. For the pilot holes, use a drill bit a little smaller than the diameter of the wood screws. Be sure to carefully center the pilot holes. Drill each pilot hole through the bottom board and about ½ inch (1.3 cm) deep into the angle-cut board. Place a piece of tape 1¼ inches (3.1 cm) from the end of the drill to guide your depth. If you make a mistake and miss the center, don't worry; just drill another hole about a half an inch (1.3 cm) away.

3. Keep the boards in place after drilling each pilot hole, and drive a 1¾-inch (4.4 cm) wood screw through the pilot hole before making the next one. Screw all four boards in place this way.

4. Line the remaining 4-foot (1.2 m) 1 x 12 piece (A) up with the edges of the 18-inch (45.7 cm) ends of the vertical boards. Drive three 1¾-inch (4.4 cm) wood screws through the top piece into each of the four dividing panels, using pilot holes as in steps 2 and 3. The top may not reach all the way down to the front as in the drawing, but it is not necessary that it does.

Securing the Lip

1. Align the 4-foot (1.2 m) length of 1 x 3 (C) with the bottom of the nesting box on the front (shorter) side.

2. Attach with two 1¾-inch (4.4 cm) screws on each end and one in the middle, using pilot holes as above.

Attaching the Perch

1. One board at a time, place the wider side of each 12-inch (30.5 cm) 1 x 2 piece (E) against the bottom of the nesting box, with two about 6 inches (15.2 cm) in from the sides of the nesting box and one in the middle. Let the perch support stick out 4 inches (10.2 cm) from the bottom of the nesting box.

2. Screw each perch support to the bottom of the nesting box with two 1¼-inch (3.2 cm) screws, using pilot holes as above.

3. Place the 4-foot (1.2 m) 1 x 2 piece (D) on top of the supports, flush with the protruding ends. Connect with one 1¼-inch (3.2 cm) screw for each support, using pilot holes as above.

Fastening to the Coop Wall

1. Attach the nesting box to the side of your coop using four metal angle braces attached with ¾-inch (1.9 cm) screws. Placement will depend on what works best with your particular coop. Sturdy 2-inch (5 cm) metal angle braces with two screws for each side of the angle should be fine to safely secure the boxes.

2. You won't need to put a back onto your nesting box if it is up against the wall. If this is not the case, however, simply attach a 4-foot x 18-inch (1.2 m x 45.7 cm) sheet of plywood (F) to the back with screws before you attach the box to the wall.

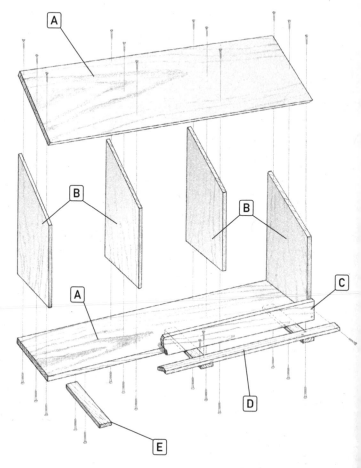

BUILDING A CHICKEN TRACTOR

You may find that mobile chicken housing suits your needs better than a permanent housing structure. Using materials and supplies readily available from any lumberyard, it is possible to custom-build a mobile housing unit for your flock. You may even be able to substitute salvaged or re-purposed materials for new wood. These plans will house four bantams, three standard-size chickens, or two heavy breeds.

CUT LIST & SUPPLIES

¼-inch (6 mm) Plywood

Number of Parts	Dimensions	Description	Comments	Ref.
8	2' x 2' (61 x 61 cm)	Four henhouse sides, henhouse roof, feed access flap, and two pitched roof panels		A
1	2' x 20½" (61 x 52.1 cm)	Henhouse floor		B

Solid wood (pine or other softwood)

Number of parts	Dimensions	Description	Comments	Ref.
2	2 x 4, 8' (2.4 m) long	Bottom sides of run frame	Make a 45° angle cut at one end of each. Also, see the **IMPORTANT NOTE** at the end of the chart.	C
2	2 x 4, 8' (2.4 m) long	Top sides of run frame		D
4	2 x 4, 4' (1.2 m) long	Vertical henhouse supports		E
2	2 x 4, 23½" (59.7 cm) long	Vertical run supports, handle-end		F
5	2 x 4, 21" (53.3 cm) long	Horizontal run supports		G
9	¼" x 1" (6 mm x 2.5 cm) wood strip, 5" (12.7 cm) long	Treads for ramp	You'll find this narrow strip in the molding section.	H
1	1 x 12 board, 15" (38.1 cm) long	Vertical side of nest box		I
1	1 x 12 board, 11" (27.9 cm) long	Horizontal side of nest box		J
1	1 x 6 board, 3' (.91 m) long	Henhouse ramp		K
1	1 x 2 board, 10½" (26.7 cm) long	Lip of nest box		L
1	1 x 2 board, 11½" (29.2 cm) long	Lip of nest box		M

IMPORTANT NOTE: As mentioned on page 55, some lumber dimensions are given here as the *nominal* dimensions they are sold by. Besides those dimensions on page 55, you'll need to know: 2 x 4 = 1½ x 3½ inches (3.8 x 8.9 cm) and 1 x 6 = ¾ x 5½ inches (1.9 x 14 cm).

CUT LIST & SUPPLIES (continued)

Other materials and supplies

Number of Parts	Dimensions	Description	Comments	Ref.
	2" x 25' (5 cm x 7.6 m)	Galvanized hardware cloth	1/2" (1.3 cm) mesh	
1	21" (53.3 cm)	1 1/2" (3.8 cm) diameter wooden dowel		
2	2" (5 cm)	Zinc-plated strap hinges		
4	2" (5 cm)	Zinc-plated hinges		
2	6" (15.2 cm)	Shelf brackets		
11	1 1/2" (3.8 cm)	Corner braces		
6	4" (10.2 cm)	Corner braces		
1		Handle or knob	Any style	
2	2 1/2" (6.4 cm)	Barrel bolts		
	1 1/4" (3.2 cm)	Exterior screws		
	2 1/2" (6.4 cm)	Exterior screws		
	1/2" (1.3 cm)	Wood screws		
120	3/16" x 1 1/4" (5 mm x 3.2 cm)	Fender washers, zinc-plated		
	4' (1.2 m)	1/2"-thick (1.3 cm) nylon rope		
	3' (.91 m)	1"-thick (2.5 cm) nylon rope		
2		Eyebolts		
1		Marine winch	May substitute a similar item for tying back rope	
		Paint	Spray or can	
1 bundle		Cedar shingles	For roof	
2	10" (25.4 cm) diameter	Wheels		
2	6 to 7" (15.2 to 17.8 cm)	Bolts		
2		Nuts	Sized to fit bolts	

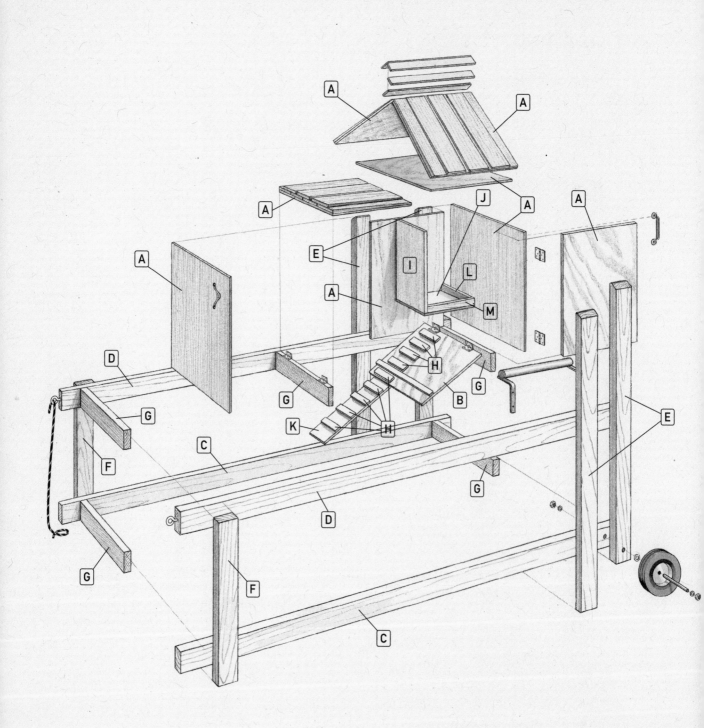

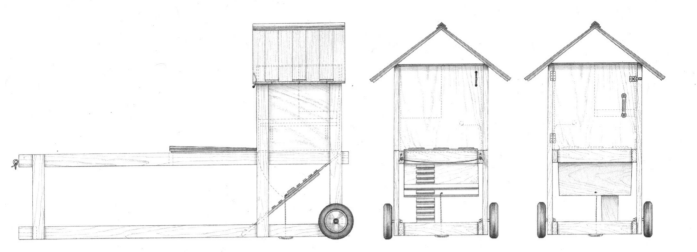

Making the Run

1. You will start by framing the run. On a flat surface, lay down the two 8-foot (2.4 m) lengths of 2 x 4 with the 45° angle cut (C). Vertically attach the two 23½-inch (59.7 cm) lengths of 2 x 4 (F) 1 inch (2.5 cm) in from each uncut end of the boards. On the opposite end, attach two 4-foot (1.2 m) 2 x 4 pieces (E) 1 inch (2.5 cm) in from the 45° angle. *Note:* Use 2 ½-inch (6.4 cm) screws for attaching two 2 x 4 pieces together.

2. Attach an 8-foot (2.4 m) length of 2 x 4 (D) to each top outside end to create a box. Attach the two remaining 4-foot (1.2 m) 2 x 4 pieces (E) 18½ inches (47 cm) apart from the other two 4-foot (1.2 m) 2 x 4 pieces first attached. Using the 1½-inch (3.8 cm) corner braces, horizontally attach two of the 21-inch (53 cm) 2 x 4 pieces (G) at each end of the tractor. One of the 21-inch (53.3 cm) pieces will go at the top of each end of the frame and two will go at the bottom of each end.

3. Next you'll attach hardware cloth inside the run. Using the washers and 1¼-inch (3.2 cm) screws, attach the hardware cloth to the interior of the frame, forming walls. Hardware cloth on the top will be attached later. Space the washers and screws out approximately every 4 inches (10.2 cm).

Building the Henhouse

1. Now you'll begin enclosing the henhouse. Center and attach two 2-foot-square (61 x 61 cm) sheets of plywood (A) between the four 4-foot (1.2 m) 2 x 4 pieces (E), creating east and west sides of the henhouse.

2. Attach the 3-foot (.91 m) 1 x 6 piece (K) to the 2-foot x 20½-inch (61 x 52.1 cm) sheet of plywood (B). Position it 2 inches (5 cm) from the 2-inch (5 cm) edge. This is the henhouse floor.

3. Next, attach the floor to the top 21-inch (53.3 cm) 2 x 4 piece (G) using 2-inch-wide (5 cm) hinges (not the strap hinges). Now you'll put treads on the 3-foot (.91 m) 1 x 6 piece. Using your ½-inch (1.3 cm) wood screws, attach the 5-inch (12.7 cm) ¼ x 1 (6 mm x 2.5 cm) pieces (H) every few inches. Once you've moved up the 3-foot (.91 m) 1 x 6 piece (K), continue attaching 5-inch (12.7 cm) pieces up the floor.

4. Using 1¼-inch (3.2 cm) screws, center and attach one 2-foot-square (61 x 61 cm) sheet of plywood (A) between the 4-foot (1.2 m) 2 x 4 pieces (E), creating a third side to the henhouse.

5. Next you will make the trap door pulley. Using a ½-inch (1.3 cm) spade bit, drill a hole halfway across the bottom edge of the floor panel, opposite the hinged edge. Drill the hole completely through the wood, and place it about ½ inch (1.3 cm) out from the third side you just attached. Tie a knot in the end of the ½-inch-thick (1.3 cm) rope and thread through the hole you just cut out. Attach the marine winch to either side of the 4-foot (1.2 m) 2 x 4 pieces (E) flanking this side.

6. Now you will attach the feed access flap. Lay one 2-foot-square (61 x 61 cm) sheet of plywood (A) down on the top of the run frame, placing it about ⅛ inch (3 mm) from the third side of the henhouse. This is will be your flap for accessing the feeder and waterer in the run.

7. Mark out where the rope would come up, and cut out a notch in the plywood for the rope to fit up through. Now, wrap the rope coming up from the floor panel around the marine winch until the floor is tightly raised.

8. After you have cut out the notch, reposition the plywood on top of the run and attach it to the frame with the two strap hinges. Situate the last remaining 21-inch (53.3 cm) piece of 2 x 4 (G) on the ceiling in the center of the run, making sure the edge meets up with the hinge edge of the feed access flap.

9. Next, you will put on the henhouse roof. Attach four 4-inch (10.2 cm) corner braces to the top of each 4-foot (1.2 m) 2 x 4 piece (E), framing the henhouse. Secure one 2-foot-square (61 x 61 cm) sheet of plywood (A) to the top of the braces. This will be the interior roof.

10. Using a ⅛-inch (3 mm) spade bit, drill about 10 holes in the roof. These will serve as ventilation.

11. Now you will install the nesting boxes. Attach the 11-inch (27.9 cm) (J) and the 15-inch (38.1 cm) (I) pieces of 1 x 12 to each other to form an L-shape. The 15-inch (38.1 cm) board should be vertical, and the 11-inch (27.9 cm) board will run horizontally, such that the 11-inch (27.9 cm) side of the L-shape will be the floor.

12. Inside the henhouse, secure this L-shape to the wall with two 4-inch (10.2 cm) angle braces using ½-inch (1.3 cm) screws. The angle braces will be on the bottom of the box. Attach the top of the 15-inch (38.1 cm) 1 x 12 (I) to the roof using a 1½-inch (3.8 cm) corner brace.

13. Secure the 10½-inch (26.7 cm) piece (L) and the 11½-inch (29.2 cm) piece (M) of 1 x 2 around the bottom of the nesting box, forming a lip that will prevent eggs from sliding out.

14. Next up is attaching the perch. Inside the henhouse, on the wall opposite from the nesting box, secure the two 6-inch (15.2 cm) shelf brackets using ½-inch (1.3 cm) screws. Space

the brackets far enough apart to accommodate the 21-inch (53.3 cm) dowel. Attach the dowel to the brackets through the top bracket hole using ¾-inch (1.9 cm) screws.

15. Now you will put on the henhouse door. Attach the remaining 2-foot-square (61 x 61 cm) sheet of plywood (A) with 2-inch (5 cm) hinges to the opening in front of the nesting box. Position the hinges on the left side of the sheet. This is your door for accessing eggs from the nesting box.

16. Secure a handle or knob to the outside of the door, on the right-hand side. Place barrel bolts at the top and bottom of the door, flanking the handle. These will help predator-proof the henhouse door.

Finishing Touches

1. To finish up, close off the run. Secure hardware cloth to the remaining open section of run, again using washers and 1¼-inch (3.2 cm) screws.

2. Now attach the rope handle. At the end of the run opposite the henhouse, attach eyebolts into the top 8-foot (2.4 m) 2 x 4 pieces. Thread 1-inch-thick (2.5 cm) rope between the eyebolts, and tie knots at either end to secure. This is the handle for pulling your chicken tractor around your yard.

3. Next comes the henhouse roof. Attach the two remaining 2-foot-square (61 x 61 cm) sheets of plywood (A) with the last two 1½-inch (3.8 cm) corner braces to form a 90° angle. Position over the top of the henhouse, and secure to the sides.

4. Screw the cedar shingles to the roof frame in an overlapping motif. Use either ½-inch (1.3 cm) or ¾-inch (1.9 cm) screws depending on the thickness of the wood, which will vary in your bundle.

5. Paint the entire tractor, which helps to both preserve it and smooth over any rough edges on the wood.

6. Finally, attach wheels to the bottom of the frame on the henhouse end of the run. Position them about ½-inch (1.3 cm) up from the bottom edge of the board, and thread nuts and bolts to secure.

Portrait of a chicken owner

Kelly and Erik

Erik and Kelly are living examples that you don't have to have a rural postal code to keep chickens. Situated in a major metropolitan area, they have been looking over their small flock of four for just over a year now. In addition to savory eggs, they have delighted in **"the usual pecking-order antics of our highly active hens."** Not only do they lack a lawn, but the couple, both home-based writers, live on a hill. Creating proper housing for their flock therefore required ingenuity and an ability to think outside the coop. Envisioning their space in terms of zones, Erik and Kelly formed three areas on their property, each serving a distinct purpose. Zone one is the henhouse, for sleeping and egg-laying. Zone two is a covered run, secured with 1/2-inch (1.3 cm) hardware cloth for deterring enterprising raccoons. Zone three is movable fencing topped with bird netting, which allows the flock to roam in a designated area, hunting, scratching, and pecking to their hearts' content.

The couple acquired the chickens from a nearby feed store partly as research for their book on urban homesteading. Erik and Kelly also wanted to **"be more connected to where our food comes from and to be a part of the cycles of life and death associated with keeping livestock."** For their part, creating housing for their flock proved that, while there is no one-size-fits-all solution, the benefits of food, fertilizer, and insect control make chickens well worth designing custom digs for. As Erik puts it, **"with all these benefits, they are the perfect creatures to integrate into our lives, homes, schools, and neighborhoods."**

Chapter 5
Feeding

Most people keep chickens in order to get something out of them, literally. In order for chickens to furnish you with well-formed eggs and tasty meat, they need to be properly fed. Age, sex, climate, and intended use all contribute to individualized feeding needs. Just as you wouldn't feed a grown man puréed carrots with peas, or give a baby a filet mignon with a glass of pinot noir, life stages in chickens determine what foods they require. Knowing who needs what and when contributes to the health and longevity of your flock.

EASY TO DIGEST

The digestive system of chickens is an incredible design. As they have no teeth (think of the adage "scarce as hen's teeth"), chickens simply gather up edibles into their beaks. From there, the tongue moves food to the esophagus and then on to the crop. The crop is a pouch in the throat, acting as essentially a holding tank of sorts, where food is packed in and stored for a time. At the end of a day of pecking, you can actually see your chicken's crop bulging. Situated in the middle of their chests, the crop will appear as a small, golf ball-sized protrusion. After it leaves the crop, food travels to the first of a chicken's two stomachs, the proventriculus, also referred to as the "true" stomach. It is here that enzymes and hydrochloric acid are introduced.

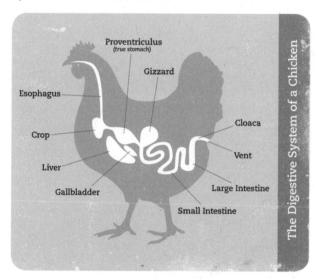

The Digestive System of a Chicken

Next, the still-undigested food moves into the second of the two stomachs, the ventriculus, better known as the gizzard. The gizzard serves as a grinding muscle, churning the food with gastric juices from the proventriculus. Any gritty material the chicken has consumed during the day helps the gizzard grind up tougher material. The resulting mass then passes to the small and large intestine where nutrients are absorbed and water is extracted. Whatever remains moves through the cloaca to be passed as waste out the vent. Amazingly, all of this occurs in a mere three to four hours. Chickens forage and consume food virtually the entirety of their waking hours. The rapid time of digestion, coupled with the prodigious amount of food consumed, accounts for the abundance of poop chickens produce.

THE RIGHT BALANCE

Chickens require a healthy balance of fats, carbohydrates, protein, vitamins, and minerals in order to fulfill their numerous biologic needs. From properly fine-tuning their internal thermostats, to providing energy, to offering the right vitamins for shell integrity, the proper ratio of dietary micro- and macronutrients is essential for optimal chicken output, not to mention a happily clucking bird.

Chickens are innate foragers. Given access to the outdoors, they will search tirelessly for bugs, worms, grass, leaves, grubs, and anything else they can get their beaks on. Their phenomenally acute vision allows them to see microscopic organisms in the soil. While birds in the wild can meet their nutritional needs themselves, domesticated fowl need assistance, even if they are permitted to forage. Commercial feeds remove the guesswork, carefully orchestrating the perfect dietary balance for you. You can further enrich their diet with supplements and treats.

DAILY RATIONS

Commercially prepared poultry feed is customized to meet chickens' daily nutritional requirements. While it is possible to formulate your own feed, it is a daunting task. Feed must include precise amounts of proteins, fats, carbohydrates, vitamins, and minerals. Such a recipe can be concocted at home, but it's a task best left to experienced chicken raisers.

Chicken feed can be purchased at your nearest feed store. Where you live largely dictates the availability of feed choices. Generally, the more rural your location, the greater access you will have to a range of ration options. Those in urban areas may have to look just outside of city limits to find a supplier. It is also possible to order feed online, no matter where you are on the map.

You can also join a buying club in your area. When I was setting up my flock, I was surprised to learn that none of the local suppliers offered organic feed. Through a friend, I became part of a buying club that makes monthly bulk purchases of organic feed from a national distributor. Each month the coordinator sends an e-mail detailing the delivery schedule. Payment is made in advance and then, on the designated day, we rendezvous in a predetermined location. The whole pick-up process feels very covert in a James-Bond-meets-the-farm sort of way. I pull up my car, give my last name, they put the goods in the trunk, and I speed away.

Commercially prepared feeds come in either mash or pellet form. Opinions vary as to which is the preferable choice. Pellets, which are, in essence, compact cylinders of food, have the advantage of reducing waste. They also help keep show chickens clean. Mash can get messy, as chickens will scatter it on the ground, and more of it will subsequently be wasted than pellets. However, mash is considered ideal for flocks kept in confined quarters. Chickens will quickly consume pellets, while it can take hours for them to eat an equivalent amount of mash. Foraging around for mash keeps them entertained and active, both of which are desirable habits for offsetting winter boredom.

SWEETENING THE POT

In addition to their daily feed, your flock will both enjoy and benefit from supplemental foods. While those with large-scale poultry operations may advise sticking to a strictly commercial

left: **Layer Crumble** *right:* **Layer Mash**

feed diet to maximize growth and production, backyard flock owners generally don't mind as much if their crew's egg output is occasionally sporadic or they take a bit longer to fatten up for table. Scratch, grit, and table scraps are all options you may want to consider giving to your winged buddies as occasional treats.

Scratch

If you want to see a whole lot of squawking and general merriment occur in your flock, sprinkle some scratch their way. Also known as "cereal feeds," scratch is simply a combination of two or more grains, in whole or coarsely cracked form. Commonly used scratch grains include corn, wheat, milo, oats, millet, rice, barley, rye, and buckwheat. Think of scratch as a treat, though, not a meal replacement. Scratch is high in fat, but low in protein. It is imperative that laying hens receive at least 16 percent total protein in their diet daily. Although hens will happily devour fat-rich grains all day, too much fat will throw off their protein consumption, which can be detrimental. Overeating scratch can also result in excess weight for your flock, which can pose its own health risks.

Many chicken tenders provide their flock a meal of mash or pellets in the morning and scatter scratch around for an afternoon delight. Toss the grains directly into their housing and chickens will peck and dig around for the scratch, fluffing up their bedding and "stomp"-composting it in the process. To help them generate and retain heat, feed your flock more scratch in the winter. Oats, which are naturally cooling when digested, make a nice substitute or addition to scratch during searing summer heat.

Grit

Whether or not you will need to offer grit to your flock depends largely on how they are housed and what they are fed. Grit is any form of small pebble or stone that chickens ingest to assist in digestion. As they have no teeth, grit offers the grinding element needed in the gizzard to break down grain and insoluble fiber. Birds kept in confinement with little to no access to the outdoors will need grit added to their diets. Flocks that are generally range-fed, consuming mostly grass and bugs, will need grit as well, to assist in breaking down plant matter. If your flock eats commercially prepared feed, either in mash or pellet form, you will not need to supplement with grit, as the grains in the feed have already been ground. Also, if your flock has access to a run or yard where stones or small pebbles may be found, adding grit is not necessary. If you are uncertain about the need to add grit or not, you can do as one natural poultry feed distributor advised me. Simply shovel up a scoop of river or streambed dirt and deposit it in your run. Don't substitute yard soil, as it may not offer the needed stones and sand. Replace with a new scoopful every four months or so, depending on your flock size. You may also simply gather up a bag of grit at your local feed store and scatter a handful to your flock or place grit in a container and allow your chickens to take what they need.

left: **Scratch** *right:* **Grit**

Table Scraps

A flock of chickens is a virtual garbage disposal machine. Chickens love to eat what we eat, as well as what's left of what we don't or won't eat. They will gladly gobble up scraps such as apple peelings, carrot tops, heels of bread, leftover cooked meat, and banana peels. Don't give them anything spoiled or moldy, onions or garlic (as they will affect the flavor of eggs), citrus fruit or peels, avocados, raw potato peels, fish, and anything too salty, fatty, or sugary. Pretty much anything else is fair game. At my house, my Ladies get a daily smattering of currants, raisins, blueberries, raspberries, wineberries, or whatever else is seasonally available and on hand. I also give them torn up bits of parsley every few days and salad or leafy greens, depending on the time of year. The chlorophyll in the greens and herbs offer nutrients called carotenoids that aid in coloring their yolks so vibrantly.

If your flock does not have access to grass in their run, consider providing them with fresh greens. Weeds pulled from your yard or vegetable garden will make a fine addition to your flock's diet. I've given mine chickweed, dandelion greens, and plantain, much to their squawking delight! If you really want to treat your flock, plant a garden just for them. Little lettuces and hardy greens are sure to please. Tossing in the occasional slug, cricket, or worm you find lying around never fails to enchant them. After picking hornworms off my tomato plants, any that I gather up go straight to the Ladies. What ensues is the chicken equivalent of a catfight, often with the hornworm as the prize in a game of tug-of-war!

WAY TO GROW!

Nutritional needs for chickens will vary throughout their lifetimes. There are five general feeding classes: chicks, broilers, pullets, layers, and breeders. Commercial feed, also known as "ration," is available for each class. Chicks and their feeding requirements will be discussed in detail on pages 85 to 86.

Broiler feed is high in protein, making up 18 to 22 percent of the feed's volume. This expedites growth, enabling chickens to become table ready quickly. Most birds intended for table are fed a broiler "starter/grower" ration until six weeks of age or so, and then transitioned to a "finisher" ration until slaughter.

Pullet feed contains no calcium and is lower in protein than chick feed. This is to prevent the birds from maturing too quickly, thereby bringing on early laying. Early maturation can result in smaller eggs and potential internal injury to your hens.

Once they are ready, around 18 to 20 weeks of age, pullets graduate to "big girl" food, known as layer rations, which provides sufficient calcium for eggshells to properly harden. Layer feed should be composed of 16 to 18 percent protein, the upper end of the percentage being reserved for birds in particularly warm weather that may be inclined to consume less food.

Breeder feed contains extra protein and vitamins for chicken mommas-to-be. It is of the utmost importance that hens destined for breeding be given appropriately formulated feed. Layer feed contains neither adequate protein

Kitchen scraps and fresh herbs provide additional nutrients to your flock.

nor the proper amounts of vitamins and minerals needed for developing fertile eggs. Experts recommend starting breeder rations two to four weeks before you plan on beginning hatching. In the event that you are unable to locate breeder feed, add a handful of dog or cat food two or three times per week to layer rations, starting six weeks before you intend to gather eggs for hatching. You will also need to incorporate a powdered vitamin and mineral supplement into the breeding hen's water.

WATER, WATER, EVERYWHERE!

Chickens need access to clean water at all times. Much like humans, chickens are composed mostly of water. Their bodies as well as their eggs depend on a continual supply of water. Deprived of adequate water for even 24 hours, a chicken will begin to suffer. Any longer and they may never fully recover. While they may not take large gulps at one time, chickens do drink copiously throughout the day, averaging around one to two cups (240 to 480 mL) per chicken. It is therefore essential to stay on top of both the availability and cleanliness of the water you offer your birds.

In warm climates and during the hottest calendar months, refresh your flock's water supply as necessary. They will consume considerably more fluid when the mercury rises. When the temperature drops, it is essential that you prevent the water supply from freezing. This can be done either by changing out waterers as needed if it's below freezing or by installing a heated waterer. Chickens are notoriously finicky, so be certain their water doesn't become too hot or too cold. Between 50° and 55°F (10° and 13°C) would be their preference. Like Goldilocks, they like things to be just right.

FEEDING FRENZY

There are two methods commonly used for feeding chickens: free choice or by hand, also referred to as restricted feeding. In free-choice feeding, the chickens themselves call the shots on when to eat and how much to consume

PORTION CONTROL

Just how much feed should you dole out per meal? That depends on the age of your chickens, their breed, and the time of year. Smaller chickens and young birds will eat less food than full-grown, heavy breeds. All chickens need increased portions during cold weather. Anywhere between 4 and 6 ounces (112 and 168 g) of feed per day is recommended for adult birds. Again, that amount will vary depending on age, climate, bird size, whether they are laying, whether they have opportunity to forage, and whether they are molting or broody. If you are looking to reduce feed costs and have a suitable location for free-ranging, allow them to forage liberally. No matter what, remember that it is very important your chickens get enough to eat. Underfed chickens won't grow properly, and their ability to lay well will be compromised.

At daily mealtimes, watch to see how quickly the food is disappearing. If they gobble it up like it's their last meal ever, they need more food. Have their droppings examined by a veterinarian to rule out worms or parasites. If the amount of food you offered in the morning looks to be untouched when you return for the evening meal, your birds might not like their feed or you may be providing more than is necessary. Check to be sure it is still fresh and not comprised of mostly vitamin and mineral dust. Sprinkle in some table scraps as temptation to rev up their appetites.

per feeding session. In restricted feeding, your flock is fed a limited amount, for only a brief period of time. Each scenario has its own set of pros and cons.

Free-choice feeding makes it more likely that all members of the flock will fill their gizzards. The constant availability of feed allows those lower in the pecking order to feed when higher-status birds may not be around. This feeding style also has the advantage of enabling chicken owners to fill the hopper less often, maybe even have a weekend away if your chickens' housing is predator-proof and self-regulating. On the other hand, chickens waste considerably more rations when they are allowed free-choice feeding. They may also overeat and put on additional unnecessary weight.

Restricted feeding, or feeding by hand, ensures the least amount of feed will be wasted. It is also the preferred method for those keeping show birds, as it keeps them cleaner and increases their enthusiasm for human interaction. Conversely, it is time-consuming and may be prohibitively so for those with busy mornings, like getting to work on time or getting the kids fed and off to school at the break of dawn. Restricted feeding may prevent lower-status birds from getting as much food as they need, as bossy hens tend to hog the feeder. If you opt for a restricted feeding program, you may need to purchase more feeders so that everyone can eat to satiety at the same time.

FEED YOUR CHICKENS WELL

The food and water you provide to your chickens will need to go in containers suitable to their needs and tendencies, which is to say, their inclination to make enormous messes. Specially designed feeders and waterers are commercially available. If you can't swing the cost of feeders or waterers, great ideas for homemade versions can be found online.

Chow Time

One common feeder design consists of either galvanized or plastic tubes, suspended off the ground by chain. Another option is metal or wood troughs that rest on the ground. Whatever you choose, you want to be certain that it discourages waste, prevents your birds from sullying the contents with chicken poop, and is easy for you to both fill and clean. Keep your feeder in a covered area outdoors, just outside the coop. This keeps the feeder out of rain and snow and encourages your crew to venture outside and dig for bugs and other beneficial creepy crawlies.

Galvanized tubes can be handy if you want a weekend away, as they can be filled with several pounds worth of feed at a time. They are also relatively easy to clean, which you should do as often as monthly, or sooner if the feeder is visibly grungy. You will want to suspend tube feeders from the ceiling so that the bottom of the feeder is at the same height as your chickens' backs. Alternatively, if you are lacking somewhere to hang

the tube from, simply place cement blocks under the feeder until it is at the same height as your birds' backs. This aids in keeping the food in the feeder instead of all over the ground and keeps any bedding material out.

Trough feeders should never be filled more than half full, or your flock will waste just as much as they eat. Well-constructed trough feeders, like tube feeders, should keep your chickens from stepping into them or roosting on them. Most are designed to be height-adjustable for lowering and raising based on your flock's age and size. Like tube feeders, they will need to be cleaned regularly to keep the feed hygienic.

Thirst Quenchers

Waterers are available for purchase in quart, and 1, 5, and 10-gallon (1 to 38 L) sizes. They are constructed of either metal or plastic. What the metal options add in weight, they more than make up for in durability. Plastic feeders are more likely to crack or break, but they will be lighter to carry if you intend to fill them somewhere other than at the coop. Like feeders, waterers should be suspended just to the level of your chicken's backs.

I have a 5-gallon (19 L) waterer that lives permanently with the chickens. Each morning, as I gather up the Ladies' feed and daily treats, I fill up a plastic plant watering can and schlep everything up to the chickens with my faithful dogs in

Organic

The term "organic" refers to an agricultural practice that preserves and replenishes soil vitality without the addition of any artificial or conventional pesticides or fertilizers. Furthermore, food that is organically grown must be done so without the use of antibiotics, synthetic hormones, genetic engineering, sewage sludge, or irradiation.

When deliberating feed options for your flock, consider organic feed. Chicken feed that is organically produced is made with grains that have not been treated with chemical herbicides, insecticides, or fertilizers. Chickens fed exclusively organic feed will naturally then provide organic eggs and meat. Evidence suggests that organically grown foods are safer and more nutritionally rich than conventionally grown foods. Organic foods are generally a bit more costly than those produced conventionally. Consider your options, read up on conventional versus organic, and decide what is best for you, your family, and your budget.

tow. If the day's heat merits it, or I worry the water will freeze within several hours, I'll bring them more later in the day. This way, I don't have to lug the whole waterer with me up to the house each time.

There are a number of devices available for giving your chickens a continuous supply of water. Automatic waterers can be found at some feed stores and at many online poultry suppliers. Electric warmers are available for those where winter weather is particularly severe. A low-tech trick is to keep two waterers and change them out during the day, allowing the frozen waterer to thaw indoors before exchanging again. Whichever type of waterer you choose, remember to keep it clean, scrubbing it as often as needed.

FRESHNESS IN STORE

It is important to keep your chickens' feed fresh. When purchasing, only buy what you can go through in two to four weeks. This will help prevent spoilage as well as maintain the efficacy of added vitamins, minerals, and supplements. Proper storage is just as important. If storing the feed indoors, place it in an airtight plastic container. Outdoors, use a galvanized steel trash bin, as rodents will often chew right through plastic containers. Secure the lid of the metal pail with a bungee cord to keep feed fresh and pesky varmints out.

Any 10-gallon (38 L) plastic or metal trash can will hold 50 pounds (22.7 kg) of feed, which is terribly convenient, as feed usually comes in 50-pound bags. Be sure the feed remains free of moisture and out of direct sunlight at all times. If kept outdoors, your feed storage bin will need to be protected from rain and snow. Make sure to use up all of the feed in one bag before adding the contents of a new bag to your feed container. If you have just a tiny portion of older feed left, scoop it out of the bin first and place it at the top of the chickens' feed for that day.

> TIP:
>
> Refer to the Chicken Care Checklist on page 128 for reminders of daily, weekly, monthly, and annual chicken chores.

Portrait of a chicken owner

Zev

A man of many hats, Zev splits his time between a 27-acre (10.9 hectares) rural mountain farm, where home for him is a dome-shaped transportable structure called a yome and a communal house in the suburbs of a midsized city. In addition to two living situations, he also runs two businesses. Zev divides his time between running a permaculture land-use consulting business and coordinating all operations on the farm, including labor, project logistics, land-sharing arrangements, and developing and implementing a master plan for recovering pasture land. Permaculture is a term used to describe lasting agricultural systems that mimic nature (see Permaculture sidebar on page 18). In permaculture-designed systems, organisms synergistically complement one another, producing a mutually beneficial, symbiotic relationship. It's essentially a win-win situation for everyone, and everything.

Without question, chickens can be an integral part of a permaculture system. Zev explains **"small animals convert plants into animal protein and fat much more efficiently than large animals do (sometimes over 10 times as efficiently), and they do it with less impact on ecosystems."** On the farm, he oversees a rotating flock of between 35 and 45 chickens and has been developing a long-term plan for feeding his chickens exclusively from crops grown on the property. As chickens are primarily insectivores, and herbivores secondarily, Zev is **"working on systems for providing our flock with grubs, seeds, greens, and fruits to eat."** He hopes to eventually have a chicken "food web" wherein his flock's diet is entirely based on bugs and crops living and growing where they are housed.

Chapter 6
Hatching Eggs

Whether you are just setting up your flock or simply reach a point where you are ready to expand, a time will come when you may consider hatching chicks yourself. This might be done initially to establish a strong, healthy flock, or perhaps as a science lesson for young children, or even simply by virtue of necessity as aging layers decrease their output. Whatever your particular motivation, consider in advance the essential requirements of successfully hatching eggs to determine if this is the best route for you. Hatching eggs is not an easy feat for the novice. If you have the time and constitution for it, though, the reward of watching an emerging chick peck its way into the world is unparalleled.

OBTAINING FERTILE EGGS

A rooster is required for fertilizing eggs. Necessarily, without a rooster nearby, any eggs your hens lay will not be fertile. In that case, there are several options available for obtaining fertile eggs. You can either take eggs from a hen in your area that is laying fertile eggs, or you can order freshly laid, fertilized eggs through the mail. To locate a hen in your area with fertile eggs, contact your local feed store, look in the classified section of your local paper, or search online for farmers and chicken enthusiasts in your area.

If you do keep a rooster as a permanent resident of your flock, your eggs are guaranteed to be fertile. If you live in an area where roosters are allowed but do not keep one, consider renting a "stud." In this scenario, a rooster of guaranteed quality breeding stock is housed with your layers for a period of time, usually a month, before he is returned to his owner. This enables you to have the benefits of home-hatched fertile eggs without all the less-desirable "extras" that come with full-time rooster ownership. Make sure that your rooster comes from healthy stock conforming to breed standards, especially if you ever intend to show your birds. Roosters with flawed traits, whether a weak constitution or twisted toes, may pass them on to their offspring. If you plan to borrow a rooster, take a good look to verify that he will be a suitable papa for your Ladies' chicks. He should be neither too old nor too young, but in active breeding age.

EXPIRATION DATE

Fertile eggs remain viable for a finite period of time, and only if certain conditions are met. After eggs are collected, incubation is advised within six days, although they may be kept for up to two weeks. The manner in which your eggs will be hatched also factors into their storage time. In general, those that are destined for artificial incubation should be stored no longer than a week, while those a broody hen will sit on can be stored for up to 10 days. The longer fertile eggs are stored, the time it takes for them to hatch increases, while their viability decreases.

Temperature is another important variable when storing eggs destined for hatching. The ideal temperature for stored eggs is 55°F (13°C). The eggs should be kept out of direct sunlight in a cool, relatively dry place. Do not put the eggs in a refrigerator. Too much humidity can create conditions conducive to mold and bacteria growth. Too dry an environment can cause moisture to evaporate through the shell. You can store your pre-incubation eggs in a regular egg carton. Place them pointy-end down and tilted a bit to one side. Each day, you will need to tilt the eggs in the opposite direction. You can easily do this by using a small paperback book to prop up one side of the carton, then simply move it to the other side of the carton around midday.

IT'S THE TIME OF THE SEASON

Time of year is important to take into consideration with hatching eggs. Left to their own devices, hens lay fertile eggs in early spring, when daylight hours are increasing. Hatching in spring is ideal for newborn chicks, as the cooler weather keeps germs and parasites that thrive in warmer weather at bay, allowing their fledgling immune systems to develop. Chicks born in this sort of climate are generally stronger and healthier than those born in the heat of summer.

Decide what time of the year to hatch based on your intentions. If your purpose is to have a supply of eggs, chicks hatched in February and March are the way to go. They will begin laying in late summer, while there are still long daylight hours, and continue laying through the fall and for the following year, although production will wane during days with reduced daylight. Chicks hatched during the winter will lay during the summer months as well, but will most likely molt in autumn and may not resume laying until the next spring. If you are interested in hatching chicks to be ready for show, be mindful that it will take standard breeds eight to 10 months to mature, while bantams will be ready to show in six to seven months.

Mother Hens

Some chicken breeds have had broodiness bred out of them. Others are known to be reliable brooders. While the following designations will always have their occasional exceptions, they are fairly trustworthy.

Consistently Broody

Ameraucana, Araucana, Australorp, Belgian d'Uccle, Brahma, Buckeye, Chantecler, Cochin, Cubalaya, Dominique, Java, Langshan, Old English Game, Orpington, Silkie, Sumatra, Sussex, Wyandotte

Occasionally Broody

Aseel, Dorking, New Hampshire Red, Plymouth Rock, Rhode Island Red, Shamo

Rarely Broody

Ancona, Andalusian, Campine, Cornish, Hamburg, Houdan, hybrid layers, Jersey Giant, Leghorn, Malay, Minorca, Polish, Spanish White Face

NEST EGGS

Eggs may be incubated either naturally, under a hen, or artificially, in an incubator. To successfully hatch eggs naturally, you will need to either obtain a broody hen or find the best brooder in your flock. While relying on a hen for incubation puts you in the position of waiting for a hen in your flock to

go broody, this method is the most fail-safe way to supply the best conditions for chick growth. The hen will take care of all the physical requirements eggs need to best hatch, without any effort on your part other than meeting the needs of your brooder, which are few. Top of that list are special breeder rations, begun about two to four weeks before you want to start hatching. Breeder rations contain supplemental protein, vitamins, and minerals, which will improve the hatchability rate of your eggs.

Many of the best breeds of laying chickens had broodiness bred out of them long ago. Once a hen gathers a clutch of eggs and starts sitting, the hormone prolactin is released by her pituitary gland. The release of prolactin provokes the cessation of laying. You can see why commercial egg producers would want to breed out the broodiness trait, as a three-week hiatus from egg laying significantly interferes with the flow of egg income. Consult breed profiles to make sure you have a breed known for broodiness before attempting natural incubation.

Even among breeds with a propensity toward broodiness, some birds are broodier than others. Do any of your hens stay in the nesting box for lengthy stints while the others peck away outside? You may have a broody hen in your flock. Hanging out in the nesting box is not conclusive proof, however. In order to determine for certain who you're dealing with, attempt to remove any eggs under your hen. If she moves off with little fuss, you don't have a brooder. If she tries to peck you, makes growling sounds, puffs out her feathers, or otherwise acts ticked off, you might be dealing with a proud mama. You can either wait for eggs to build up in the nest or place fake eggs made of either wood or plastic in the nesting boxes and see who takes to the throne. Before committing fertile eggs to your suspected brooder, place a few fake eggs under her for the night. If she is still there the next day, looking like she is ready to start hanging curtains and painting the nursery, you've got your broody hen! Go ahead then and replace the fake eggs with fertile ones.

Broody hens will need their nests isolated from the rest of the flock. During her three-week sitting period, other hens may bully her or even attempt to take over her nest when she is out eating or doing her business. After the chicks are born, they may become a target of their flock mates, getting stomped on or eaten by them. Newly hatched chicks are considered scrumptious, easy morsels by predators from above and below. Keeping your broody hen, her eggs, and the eventual chicks safe should be a top priority.

An ideal nest setting is somewhat dark, well-ventilated, predator-proof, and protected from the elements. It needs to be about 15 inches (38.1 cm) square and 16 inches (40.6 cm)

A hen exhibiting broodiness

high. A cardboard box with 1-inch (2.5 cm) slits cut in the roof for ventilation would be fine. After you are certain your hen will sit, you can remove one side of the box so she can come and go as she desires.

Cedar shavings make an ideal nest material, as they provide both a cushion for the eggs and help keep mites and lice away. It is especially important to keep these parasites away, as they can consume enough blood to easily kill newly hatched chicks. If your hen initially sets up her nest in the henhouse or in some other unsuitable area, you will need to relocate her. Do this at night, when she is calmer, and be sure to wear gloves, as a disturbed brooder can take opportunistic pecks, and hard pecks at that. If you have more than one brooder at the same time, each hen will need her own nesting box in separate and distinct areas.

Once she begins sitting, expect your hen to eat, drink, and move very little, usually no more than 30 minutes per day, sometimes less. She'll be busy feathering her nest, literally. A broody hen will pick off feathers from her breast to keep the eggs, and later, the chicks, closer to her warm body. This also allows moisture from her body to keep the eggs from drying out over the three-week sitting period. You can put small bowls of both food and water in the corners of her nest, or just outside it, to encourage her to eat. If she becomes so fixated on the task of mothering that she neglects to eat, you may need to lift her off the nest. Tempt her with scratch thrown right in front of her nest. Be sure to feel up around and under her wings before attempting to lift her, as she may have an egg tucked in there. Make sure she leaves her nest for no more than 30 minutes daily.

After day 16, don't bother her any further. Make sure her physical needs are met, and then let her go about her business while you go about yours. After the chicks begin hatching, the hen should continue to sit. If she ventures out of the nest before all the chicks have hatched, perhaps heading out to look after some curious and adventuresome first-borns, gather up the hatched chicks and care for them. Return the hen to the nest to continue sitting on the unhatched eggs, then return the hatched chicks to her once the whole clutch hatches.

INCUBATING EGGS

If you have fertile eggs, but no broody hen, you can use an incubator to hatch the chicks. While most current incubators are electric, it is still possible to come across oil- or kerosene-based models. Using an incubator for hatching eggs is a tricky dance and one that requires no small amount of skill and knowledge.

Top Models

There are essentially two major types of incubators: tabletop models, which, as their name implies, sit on tables; and chest models, which rest on the floor. For the small production or backyard chicken enthusiast, the tabletop model will more than adequately meet your needs. For those engaging in large-scale poultry production, the chest model is a more suitable option, holding up to several hundred eggs at a time. Within these two types of incubators, what further distinguishes one from another is the manner in which air is circulated. You can purchase either a forced-air or a still-air incubator. Forced-air models have built-in fans that continually circulate air, while still-air models have vents on their tops, sides, and bottom. The forced-air models are considered to be significantly more reliable. Humidity and temperature—the two absolute must-haves for proper hatching—are easier to maintain in these models. Still-air models do work, however, and are usually less costly than forced-air models; they simply require greater vigilance and oversight, as well as foreknowledge that they have reduced rates of hatchability.

Whatever model you opt for, be certain to situate it out of direct sunlight. Putting it in the sun increases the likelihood of temperature fluctuations, which can be a death knell for incubating eggs. You also want to avoid particularly drafty areas, air-conditioning, or directly in front of heat vents. The three variables that would otherwise be maintained by a broody hen and that are most necessary for successful hatches are temperature, egg turning, and humidity.

Temperature

Your incubator will need a thermometer. Forced-air models usually come equipped with a thermometer, whereas still-air models generally do not. Temperature in most still-air models should be set at 102°F (39°C) and at 99.5°F (37.5°C) in forced-air models. While temperature settings will vary by incubator model, the differences are slight. Fluctuations either way can cause death to your eggs, although raising the temperature more often poses a greater threat than lowering it. Be certain to read the manufacturer's instructions that come with your incubator to learn how to best regulate temperature in your model.

Humidity

Maintaining the proper level of humidity in your incubator is also crucial. During the incubating process, eggs lose a little weight as moisture evaporation through the shell. Keeping a humid environment helps to regulate and prevent some of that loss, allowing the eggs to hatch and the chicks to emerge from the shell free of complication. If the growing embryos receive too much humidity, they will grow too large and struggle to get free of their shells upon hatching. Too little humidity and they end up sticking to the shell inside. I can think of few things sadder than a glued-in chick.

Inside the incubator, humidity will need to be at about 55 to 60 percent from day one until day 18, when it shifts up to 70 percent. Humidity levels can be measured with a hygrometer, also called a wet bulb thermometer. Again, consult the manufacturer's instructions that came with your incubator to ensure you are maintaining humidity levels at the percentage needed during different cycles of embryonic development.

Egg Turning

Eggs must be turned during incubation in order to keep the developing embryo from sticking to the shell membrane surrounding it. In more rudimentary terms, egg turning keeps the yolk in the middle of the white inside the egg. If an egg remains in one position for too long, either the embryo can become stuck and eventually die, a crippled or deformed chick will result, or the hatching chick will not be able to properly emerge from its shell. A hen on her nest performs this task naturally about every 15 minutes, either by using her beak to roll the eggs or moving them as she shifts and resettles her weight on the clutch.

In the absence of a hen, you will need to turn the eggs yourself, at least three times daily, more if possible. If you are turning them yourself, turn them as early as possible in the morning and as late as possible in the evening, so

An egg turning tray

that the time lag between turns is as short as possible. The eggs will be turned from side to side, not top to bottom. Think of the way a hen would move eggs under her on a nest. In order to keep track of which eggs you have turned, mark the tops and bottoms with a pencil with an X on one end and an O on the other.

If you will not be on hand as often as required to turn your eggs, then you will need an automatic turning device. Many incubator models come with automatic egg turners that move the eggs at regular intervals, or you can buy a separate egg turner accessory.

SHINE A LIGHT

If you elect to use an incubator, you should check your eggs weekly to be sure the embryos are developing properly and remove any that have spoiled or are maturing incorrectly. Spoiled eggs give off harmful gases that can be breathed in by the other healthy eggs. Candling will tell you which eggs should stay and which should go.

To candle an egg, a bright light is shined through to get some sense of what might be happening inside. In the past, candling was literal, with eggs being held up to an actual candle to check for fertility. Today, electric candling devices can be purchased from poultry suppliers or homemade.

When candling, you are looking for two things: the existence of a living, viable embryo, and the position of the airspace inside, which is a good indicator of humidity levels. The procedure causes no harm to the embryo as long as you do it quickly and not at all after day 18. First the eggs must be removed from the incubator and allowed to come to room temperature. This is fine at the end of weeks one and two, but not thereafter. In a darkened room, hold the eggs one at a time in front of a bright light shone through an egg-shaped hole in a box. Large commercial production facilities use a horizontal light box that can examine numerous eggs at once. For small-time chicken tenders, a small

box with a 40- to 75-watt bulb inside is sufficient. I've also read about candling by means of a slide projector or flashlight affixed to a toilet paper tube!

Fertility Awareness

Whether you purchase a candling device or improvise your own, your objective is to have light shine through the egg strongly enough that you can gather valuable information. The first bit of information is whether or not you have a fertile egg on your hands. When you shine a light through the egg, you will see one of three basic forms inside. A small black spot with spiderlike radiations out from it indicates a fertile egg. Eggs that look empty with only a light shadow from the yolk visible are eggs that were never fertile. Those eggs containing a darker shadow with a barely discernible blood ring encircling the middle are embryos that have died. Remove infertile or spoiled eggs from the incubator to avoid contamination.

Airspace Indications

The other thing you are looking for when candling is the position of the airspace inside the egg. Examining the airspace is a good way to determine humidity levels inside your eggs. As evaporation causes moisture to flow out of an egg's shell, the contents inside will decrease in size, which in turn causes the airspace to become larger. Use the diagram to see if the airspace in your eggs is similar in size relative to their growing stage. If the airspace inside looks smaller than the diagram, try decreasing humidity in your incubator. If the airspaces look proportionately larger than those in the diagram, you will need to increase humidity levels inside. Getting good at doing this comes with time and practice, so if you have difficulty at first, don't abandon hope for eventual airspace-recognition expertise!

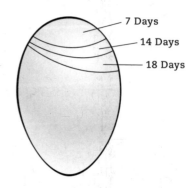

Portrait of a chicken owner

Jenny and friend

Jenny is a woman on a mission. Make that a woman on a mitzvah, the Hebrew term associated with performing a good deed. After a run-in with the city over her small urban flock of hens, Jenny was determined that keeping chickens is a right that should be extended to all residents in her community. A science museum educator and Hebrew teacher, this seemed to her like the perfect opportunity to offer a mitzvah to her fellow citizens. For the past several years, she has campaigned tirelessly to have the ordinance regarding keeping chickens within city limits changed. Additionally, she has coordinated coop tours, had T-shirts created, engaged in outreach programs, and even marched in chicken suits with other volunteers in her city's annual holiday parade.

Her efforts have benefited her children, in addition to the wider community. As she puts it, **"I taught my children that when it is important, you never give up. Our dogged persistence resulted in our having a special permit issued. We now serve as a model to our neighbors as well as a little petting zoo."** Not only did keeping chickens offer a number of lessons in biology to her children, it also instilled in them a sense of pride and responsibility, not to mention bragging rights. Keeping chickens in the front yard of their tightly knit neighborhood gave both her and her family "instant cool" in the eyes of their peers, especially during the childrens' teen years.

Jenny has been keeping chickens for 14 years in her 1920s home. Many, many lessons have been learned along the way. Once, when one of her hens became broody, she solicited fertile eggs from a nearby nature center, since, lacking a rooster, her hens had no fertile eggs of their own. She left the hen to her own devices, returning from a camping trip after the requisite 21 days needed for hatching. As the chicks began to pip, their foster mother became enraged at the interlopers gathering underneath her and began attacking them. Jenny rescued the chicks, brooded them indoors until it seemed they were capable of living in the coop, and released them, only to have one turn out to be a rooster (not allowed in her neighborhood, even under special permit) and the other fall prey to an opportunistic raccoon.

No matter the learning curves with her flock, it's worth it for Jenny. **"They are so very delightful, entertaining, and stunningly beautiful, with complex personalities. They know the sound of my car when it pulls up and rush to the fence to greet me. They are a pleasure and a blessing."**

Chapter 7
Raising Chicks

Few things rival the cuteness factor presented by newly hatched chicks. Covered in downy fuzz, peeping nonstop, and weighing about as much as a mushroom, chicks can warm the heart of young and old alike. Like any newborn, they are needy little creatures. In order to ensure your cute fuzzies properly develop into healthy hens and roosters, watchful and informed vigilance will be necessary on your part.

FEATHERY MOMS OR TIN HENS?

The manner in which you will rear your chicks depends largely on whether they were hatched under a broody hen or artificially incubated. Housing for chicks, with or without a hen around, is referred to as a brooder. Chicks with a hen present at hatching will be largely shown what to do by her, although it's a good idea to keep an eye out to confirm that they are drinking and eating.

You will need to provide food and water in chick-sized containers (discussion about these needs forthcoming) that the hen can-not access. She will need her own food and water, suited to her needs, and situated in locations only she can reach; for example, hung from the ceiling or placed on cement blocks or bricks.

Chicks with a hen available will be warmed by her and require no external heat source. Should they be hatched during cooler weather, you will need to protect both hen and chicks from es-pecially chilly temperatures and drafts. Keep the babies separate from the rest of the flock for the first eight weeks of their lives, as the petite size of the chicks makes them particularly vulnerable to being tramped on or pecked at.

If your chicks were hatched in an incubator, you will need to meet all of their needs, including regulating temperature, providing food and water, and keeping their living quarters clean. You can purchase complete brooder housing through poultry

catalogs or online, or you can inexpensively fashion a home-made version. In large-scale poultry operations, chicks born in an incubator are often moved to a battery. This style of brooder is made of tiers, sometimes stackable, with each tier containing its own heating coil. Batteries also have feeders and waterers built into their sides.

Those engaged in small production poultry businesses, or simply wishing to add to their backyard flock, can opt for hover, heat lamp, or incandescent bulb brooder setups. These types of brooders provide the heat that is absolutely essential to keeping young chicks alive and thriving. A hover is a large metal um-brella-like heater that hangs from the ceiling literally hovering over the chicks. Hovers can easily provide heat for 100 or more chicks. Many hover models have curtain-like panels hanging over their edges to help in draft protection and heat retention. Many farm-supply stores carry hovers; it is also easy to locate suppliers online.

Heat lamps and incandescent lightbulbs can also be used for heat in brooders. If using heat lamps, you will need to purchase an infrared bulb, either red or clear. Red bulbs are recommend-ed, as they reduce visibility, which in turn helps prevent picking, a potentially serious problem. Select a 75- to 100-watt infrared bulb. If using incandescent bulbs, anywhere between 60 and 100 watts will do, depending on the number of chicks you are brood-ing. Watch your chicks' behavior to see if the wattage you've chosen is too hot, too cool, or just right.

A homemade brooder fashioned out of a plastic storage bin

Whether you opt for an infrared or an incandescent bulb, suspend it above the brooder housing on a metal chain, not hung by its cord. Either type of bulb also needs to be fitted into a porcelain, not plastic, socket. The lamps will be left on continually in the beginning, and plastic sockets can melt from the heat. Lastly, both styles of bulbs will need to be placed in a reflector with a protective wire cover in the front. Doing so serves as a precautionary measure against fire should the lamp fall over.

CLIMATE CHANGE

Outside of food and water, nothing is of greater importance for your chicks than staying warm. Under the feathered breast of a hen, this need is met. With no hen to warm your little puffs, it will be up to you to keep your chicks toasty. Whether you use a hover, a heat lamp, or an incandescent lightbulb, initially the heat source will need to be placed 18 inches (45.7 cm) from the brooder floor. Every week, move it up about 3 inches (7.6 cm) or so, adjusting the placement depending on your chicks' behavior. If the brooder is too toasty, chicks will spread out to the sides, sometimes piling up on each other, causing suffocation. Too cool and they will pile up in the center, sometimes with the same end result. When temps are just right, the chicks will be spread out evenly across the brooder floor. Be mindful of their actions and adjust the lamp's location accordingly.

In battery or hover brooders, you can regulate the temperature with an adjustable thermostat. Measure the air temperature immediately surrounding the chicks by hanging a thermometer about 2 inches (5 cm) above the brooder floor. When they first hatch and continuing on for one week, chicks will need the brooder temperature to be about 95°F (35°C). Each week thereafter, the temperature can be decreased by 5°F (3°C) until the temperature reaches 70°F (21°C), or room temperature. Thereafter, the chicks should be able to regulate their own temperature, as their feathers will have begun coming in.

DRAFT DODGING

It will also be necessary to keep the chicks away from drafts. Situate your brooder in the least drafty place. Some choose to place their brooder setup in an outbuilding, such as an unused coop, while others place the chicks in the basement, the garage,

One home brooder setup option

in a spare room indoors, or even in the kitchen. No matter where you locate your chicks, be sure there are no obvious drafts coming in on them, either through air-conditioning vents or cracks in the basement wall. Many people opt to place the brooder on a table or some other structure that enables the chicks to be kept off the floor. Cold air falls and can chill little peeps pretty quickly.

A brooder guard or barrier can keep most drafts away. Improvise a barrier by housing the chicks in a cardboard box, a plastic storage bin, an aquarium, or even an animal feeding or watering trough. You can also simply take a large cardboard box and cut out a circle at least 1 foot (30.5 cm) in diameter. Secure the ends with duct tape, and place the cardboard circle about 2 to 3 feet (61 to 91.4 cm) away from the heat source. Replace the circle with a new, larger one as the chicks grow.

Brooder guards help chicks locate their food, water, and heat sources with greater ease when they are just getting familiar with the layout of their new home. Chilled, hungry, or dehydrated chicks can succumb to illness or worse rather quickly. Where chicks are concerned, prevention is considerably easier than cure. Have their setup ready before their arrival to truly ensure you have accounted for any possible oversights. It also goes without saying that whatever housing you use should be predator-proof, whether you are trying to keep out Ricky the raccoon or Fluffy the cat.

HOUSEKEEPING ESSENTIALS

Like any infant, chicks need housing that keeps their diminutive size in mind. A soft bed, adequate space, and somewhere to perch gently are key to helping your little ones grow up strong and healthy.

Bedding

Inside the brooder, the chicks will need some sort of bedding. Chick bedding should be like the perfect pair of winter boots: well insulated, absorbent, and nonslip. Little chickie legs need bedding that doesn't slide around, otherwise they can end up with a condition called spraddle leg, which is exactly what it sounds like. They also need bedding that keeps them warm and wicks away moisture produced by spilt water and droppings.

Ideal bedding materials include pine shavings, peat moss, ground-up corn-cobs or rice hulls, or even old cloth rags. Avoid sawdust, straw, and sand. After a month, you can transition to cedar shavings. Don't do so beforehand, as chicks, in their ever-present pecking, may ingest some, which can be toxic and may cause intestinal blockages. Put down about 4 to 5 inches (10.2 to 12.7 cm) of bedding to begin with, adding more as necessary to keep it fluffy. Stir the bedding daily and remove any moist areas, which can quickly become breeding grounds for bacteria.

Space

The amount of space you will need to provide in your brooder is mostly based on what type you are using. If you are using a battery, your chicks will need 10 square inches (64.5 cm²) each until they are two weeks old, then an additional 10 square inches (64.5 cm²) per chick every two weeks thereafter. If your brooder is a hover or heat lamp, your chicks will need 6 square inches (38.7 cm²) per chick until they are four weeks old and then 1 square foot (77.4 cm²) per chick until they are eight weeks. After they are eight weeks old, they can be moved into the "big house" with the rest of your flock, with each bird needing between 2½ and 3½ square feet (.22 and .32 m²), depending on breed and type (bantam, light, or heavy). Give them the room they need right from the start. Crowded chicks can turn into pecking, and—horror of horrors—cannibalistic chicks. Sad, but true.

Roosts

When your chicks are around three to four weeks old, they will be ready to begin roosting. To get them started, place perches low in their brooder, only a few inches off the floor. As they grow, you can place the roosts higher up. Give each chick about 4 inches (10.2 cm) worth of roost space. Since their feet are so tiny at this point, use a thin dowel until it becomes obvious the dowel size needs to be increased. Watch your chicks and see who gets the whole roosting idea and who doesn't. For the little peeps that don't seem to get it, gently place them on the roost, removing your hand only after they have gripped the dowel. They may fall off, but that's okay. Eventually, they will literally get the hang of it and never forget how it's done.

FOOD AND WATER

Whether your chicks are hatched under a warm hen, in an incubator, or come to you in a cardboard box via express mail, they will need to eat and drink fairly soon. The yolk offers nutrients to newly hatched chicks, keeping them hydrated and nourished until everyone in the nest comes pipping out of their shells. This works perfectly for chicks who are shipped just after hatching, arriving one or two days later. When the yolk is gone, your chicks will need to be introduced to food and water, in containers suited to their needs.

Chicks and Food

Baby chickens have unique nutritional needs, which are easily met by specially formulated chick feeds. Both starter broiler and starter layer feeds are available, so make your selection accordingly. Chick feeds are available in mash or pellet form.

Medication to prevent coccidiosis, the most common illness chicks face, is frequently added to chick feed. The choice to use medicated feed is up to you. In general, if your flock is small, your brooder cleaned daily, and the climate when your chicks arrive isn't sweltering, medicated starter may not be necessary. Conversely, if you are brooding large numbers of chicks, you get started in muggy, warm weather, or you just can't manage to stay on top of daily brooder cleanup, medicated may be the way to go. Chick grit, which is smaller than grit for full-grown chickens, will need to be supplied only if you offer chick scratch.

Place chick feed in troughs, if you are feeding many chicks at once, or in screw-top bases, if your flock is small. Trough feeders either have rods on top that turn in place, discouraging chicks from standing on them, or small openings for chicks to put their tiny heads into. The screw-top bases have several small circular openings and simply screw onto inverted glass jars. Both feeder styles can be picked up inexpensively either online or at your local feed or farm supply store.

A trough feeder

Chicks are notoriously messy, as well as prolifically poopy, so be certain to use feeders specific to chicks. Contamination from droppings in their food or water can make chicks very sick. Chick feeders are specially designed to prevent chicks from stepping into their feeders and have openings small enough to accommodate tiny beaks. Place the feeder no farther than 2 feet (61 cm) from the heat source in the brooder for the first week and never farther than 10 feet (3 m) until they move into the main coop.

Chicks and Water

A dehydrated chick can quickly turn into a sick or dead chick, so providing a constant supply of fresh water is essential. Chick waterers come as either galvanized or plastic screw-top bases, similar to chick feeders, or as plastic tube waterers with small

A typical chick waterer

openings. In a pinch, you can devise a homemade version by filling up a glass jar with water, placing a saucer roughly the same size on top, and flipping the whole thing over. You will need to lodge something like a toothpick between the jar and saucer so that the water can trickle out. A quart (liter) jar will meet the water needs of a small flock while a 1-gallon (3.8 L) waterer can tend 40 to 50 chicks. The lips of some chick waterers are a little deep, posing threat of drowning, so adding rocks or marbles is advised. Remember to change the water out at least once daily, keep the waterer clean, and never let the water supply run out. Like feeders, keep waterers 2 feet (61 cm) from the heat source for the first week, eventually moving them no farther than 10 feet (3 m) away until chicks are introduced to the rest of the flock.

It may be necessary at first to teach chicks how to drink, especially in the absence of a hen. Simply gather them up one by one and gently dip their beaks into the water. They'll figure it out pretty quickly, and others will catch on by watching the leaders. Some chick aficionados suggest spiking chicks' water with a chick-specific energy booster. Vitamin and electrolyte chick supplements can be purchased through feed stores and hatcheries. As a one-time treat for reviving the energy and diminishing the stress of mail-order chicks, add ¼ cup (56 g) of table sugar to 1 gallon (3.8 L) of water.

IT'S A HARD-KNOCK LIFE, FOR CHICKS

There are a number of ailments posing specific risks to chicks. Keep an eye out for potential problems including coccidiosis, pasting, and picking.

Coccidiosis

Coccidia are protozoa naturally colonizing the intestines of chicks. Exposure to coccidia occurs when chicks come in contact with their droppings. Under normal conditions, coccidia multiply slowly enough that chicks are able to develop immunity to them around 14 weeks of age. The protozoa thrives in damp, warm conditions and can rapidly proliferate in the presence of large amounts of chicken droppings or moist, manure-encrusted feeders and waterers. One of the earliest symptoms of coccidiosis is runny droppings, with bloody droppings showing up later, when the illness is pretty far along. If you have not been feeding your chicks medicated feed and these symptoms present themselves, switch to medicated feed, isolate the sick chicks, and tidy up and sanitize everything. To prevent coccidiosis from ever occurring, either use medicated feed or practice scrupulous brooder hygiene.

Pasting

Pasting (also known as "pasting up," "pasty butt," and "sticky bottom") occurs when a chick becomes stressed from travel, chilled or overheated, dehydrated, or ingests bedding material. Their bottoms subsequently become crusted over with droppings (a mental image best avoided, if there ever was one). This is a very serious condition, requiring immediate action to prevent death, should it occur. You will need to hold the chick's rear end under lukewarm running water and gently dislodge the droppings, or use a warm washcloth. It wouldn't hurt to rub a natural lubricant such as sweet almond, coconut, or Vitamin E oil on their bums afterward. They will fuss like their lives depend on it, but don't give up. Attempt to determine what caused the pasting problem to begin with, and make adjustments as necessary.

Picking

Chicks are prone to the unfortunate habit of picking, which is pecking at each other's toes and feathers, sometimes to the point of death. Usually, this is brought on by overcrowding, overheating, lack of food, too much light in the brooder, stagnant air, or insufficient protein in their food. Picking is definitely easier to prevent than to treat, so take steps in advance and curb the habit quickly as soon as you know it is happening. Remove any injured birds, clean their wounds, and apply a healing salve. It might also be worth switching to red bulbs, which impair visibility, and adding grass clippings to the brooder to keep them occupied.

Portrait of a chicken owner

Natasha

Raising chickens on her family farm in western North Carolina, Natasha is well versed in the fastidious care and diligence required to raise chicks. When not busy either freelance writing or working for a nonprofit agency that helps children serve their communities, she can be found caring for her flock of a dozen chickens. Situated on a dirt road that a family friend refers to as **"not a road, but a challenge,"** Natasha adds new members to her flock by purchasing newly hatched chicks, which she prefers to source locally, even if that limits her breed options. For those unable to secure chicks from a nearby location, large hatcheries are a worthwhile alternative. "What works about ordering from a hatchery is that the chicks arrive and subsist on the protein from the egg for up to three days," she enthuses.

Little is needed in the way of materials to successfully raise chicks, Natasha contends. Grab a metal washtub, a utility light with a strong bulb, add some chick feed and water in suitable receptacles, and you're ready! When old enough to transition out of the brooder and into the coop, make sure there are no holes large enough for chicks to wriggle through. Given the chance, she notes, they will make a quick escape, leaving you to search high and low for missing chicks. While she does keep birds for table, it is the eggs her flock provides that truly delight her. **"I like finding eggs in the morning. It is like receiving little presents, little affirmations that everything is okay."**

Chapter 8
Health and Wellness

In an ideal world, your flock would remain vibrant, content, and healthy for all the days of their lives. For many small flock owners, this will be the case. Others, however, will encounter the occasional chicken health crisis. While this chapter is by no means comprehensive, it does address the most common major and minor illnesses that may befall your feathered friends. Many of the conditions listed here are rare in small flocks, but are good to be informed of nevertheless.

A CLEAN BILL OF HEALTH

Perhaps the single most important thing you can do toward ensuring the health and safety of your flock is to stay on top of coop hygiene. Dirty chickens and dirty coops can quickly spell sick chickens, or worse. A little daily elbow grease on your part acts

as preventative care. Should something manage to make them ill anyway, having a clean coop will work in their favor toward fighting the illness off.

Remove droppings daily, or, when using the deep litter method, stir droppings and add litter if needed. Make sure food and water bowls are clean and free of droppings or other debris. Always have clean food and water available. Be sure the coop is well ventilated but free from any cold drafts. Perform a thorough coop cleaning once or twice a year, complete with changing all bedding, scrubbing roosts and nesting boxes if needed, and disinfecting the entire area. In short, do whatever you can to maintain the cleanliness and integrity of your coop.

Limit your flock's exposure to other poultry and wild birds. Only introduce new chickens you know to be disease-free and healthy. Do as much as you can to minimize stress on your flock, whether that stress comes in the form of a terrorizing family pet, exposure to loud and startling noises on a daily basis, or a child who hugs their tender bodies too tightly.

STOP, LOOK, AND LISTEN

In addition to staying on top of your daily cleaning routine, take time to listen to and look at your flock each day. Chickens are fairly consistent in their behavior. They also generally stay together in their activities; you know, "birds of a feather flock together." If a member of your crew looks listless or lethargic and isn't keeping up with everyone else's goings-on, it might spell trouble. Or it might not, but it's worth it to take the extra minute or two and examine the bird in question.

I make a point of checking feet, eyes, and sounds everyday. As they gather around the feeder for their breakfast, I squat down and make sure all feet and toes are straight and flat on the ground. Next I look them square in the eye, slowly moving my gaze across the entirety of their bodies, noting anything that might give me pause. Then, I listen. Any rasping? Any sniffling? Anything other than the standard soft clucking and "that piece of dried corn is mine, fool!" reproach overheard? Finally, I look at their droppings, checking for anything short of compact, firm plops with white peaks. So, take a daily long, hard stare and give a good listen. Some things are easy to catch early if you stay in touch with your flock.

HEALTHY CHICKEN CHECKLIST

In order to determine when something is amiss, you need to know what a healthy chicken should look and sound like. Make a mental checklist, or even write it down when you are first starting out, and give it a daily rundown as you feed your flock and gather up eggs. Catch symptoms when they first surface, before they have time to advance to a genuine health crisis. From comb to toes, here's what to look for:

Anatomy	Ideal Appearance
Comb and wattles	Glossy and plump
Eyes	Shiny, bright, absent of fluid
Nostrils	No obstructions or sounds
Breast	Full and rounded
Feathers	Smooth and slick, well groomed
Vent	Clean, a bit of moisture is fine
Droppings	Firm, white-capped
Body Weight	Appropriate for age, neither over- or underweight

Additionally, healthy chickens hold their heads and tails up high, often craning and turning their heads to follow sounds and movements.

PROPER HANDLING

You will occasionally need to pick up your chickens, whether to examine them for mites, administer medications, or simply to give (and get) a hug. Handling them in advance of this need makes them more acquainted with the whole experience and can abate the stress such handling may incur. Think of it as the poultry equivalent of taking the dog along on errands in the car so that when the time comes for the annual visit with the veterinarian, Fido doesn't suddenly decide to freak out on you.

To pick up a chicken, you can do one of several things. I'll start with what works for me. Since I visit my flock often, they are accustomed to the sight of me and gather around my feet. Actually, they sometimes cluster in so tightly that I am immobilized for a few seconds until they decide to free me from their chicken circle. In any event, because of their closeness, it's usually pretty easy for me to simply squat down behind one of them, place my right hand across her back, stretching the thumb and pinkie over each wing, and then, cradling the underside of her body with my left arm, lift her up to my chest (reverse hands if you're a lefty). You must keep your right hand over their wings to keep them from flying away. I'll usually stroke their necks and backs while I have them this way, to relax them, and tuck them a bit into the crook of my arm.

Alternatively, you can corner a chicken or drop some tasty food treats near you, and once one approaches, pick her up with both hands, tucking her under your arm with her beak toward your back. This is a helpful position if you need to examine feet or vents. Other tactics for catching less tame chickens include simply picking them up off their perches while they are resting at night (when they are very docile), gathering them up in a fishing or butterfly net, and utilizing a device called a catching hook, which is basically a long pole with a V-shaped hook at the end that grabs hold of a chicken's leg, like the hook you might use to pull limelight-hogging clowns offstage, only scaled down to chicken size.

Eventually, you will likely have to make physical contact with chickens whose time on earth has come to a close. Mortality rates in healthy flocks are about 5 percent annually. Should you suffer several deaths in close sequence, though, you might have a problem on your hands. Properly disposing of deceased chickens is extremely important. If the cause of death was attack by a predator or some other cause not related to disease, it may be unnecessary to contact the authorities. Depending on where you live, you may be able to burn the chicken in an enclosed container, such as a barrel. If your chicken met its demise from a known or suspected disease, however, you may be legally required to notify your local extension or agricultural office. In some instances, they will give you the go ahead to burn your afflicted bird. On occasion they will want to examine the bird to determine what the disease is. Several poultry diseases pose a severe enough threat as to potentially endanger the rest of your flock or the greater community as a whole. When in doubt, call up your local extension agent and find out the proper protocol for your area. Never bury a dead chicken that was sick, as an animal could dig it up, thereby exposing itself and wild and domesticated animal populations as a whole to whatever did your bird in, or disease organisms could leach into groundwater supplies.

TIME OUT

Sick or injured birds need to be separated from the rest of the flock until they heal. Other flock members sometimes even pose the greatest threat to a chicken's recovery. Injured birds can fall victim to picking and general bossing around by their flock mates. One of the best means of sequestering I have discovered is a dog crate.

On one terribly unfortunate afternoon, I'd gone out for a few hours, and upon returning found the gate to my chicken's coop open and the chickens spread out in the forest surrounding their quarters. My German Shepherd saw this situation at precisely the same moment I did and tore off after the flock, as I ran screaming behind her. She caught my Buff Orpington, appropriately named Buffy, and removed a fairly sizable portion

A dog crate makes a great portable shelter for a bird in need of time alone.

of her back-feathers, skin, the whole shebang. While my husband rounded up the dogs, I gathered up Buffy and cradled her close, talking softly to her. Chickens can weather pretty horrific injuries, but succumb much more easily to stress, so I wanted to calm her as much as possible. After pouring hydrogen peroxide over her wound, I kept her in a large dog crate for about a week, administering poultry antibiotic by beak twice daily. If you think children can be stubborn about taking their medicine, try a chicken, but that's a whole other story. For the first few days, I kept her in the henhouse all day. We threaded a dowel through the crate to provide her with a perch and gave her fresh food and water in small metal bowls. Two days later, I felt comfortable moving the crate outside during the day, removing its plastic floor so that she could peck at bugs and grass. After one week, the wound had begun to heal over, so I allowed her to rejoin her flock mates. Of course, within the first few minutes of her release, two of them pecked at her still-open wound. She replied in a high-pitched squawk that I can only imagine meant "Don't even *think* about doing that again, or I will have to take you down!"

Consider keeping a dog crate on hand, or some similar housing, for placing sick or injured birds into. A chicken tractor works well for this purpose also. Such a setup could also be suitable housing for a broody hen. Keeping housing for isolating a chicken readily available saves you the added stress of a last-minute scramble and gives your feathered friend somewhere safe to go, fast!

IN SICKNESS AND IN HEALTH

Most of the maladies that may affect your flock will come from diseases, parasites, or other flock members. While the cause of diseases and parasites is usually from an external source (oftentimes wild birds), they can also be introduced to a healthy flock when new chickens are acquired. Only buy birds as chicks from a reliable hatchery, or as older birds from a seller of outstanding repute. Sometimes, even in the best of circumstances, birds that are healthy in one flock will cause illness in another. Keep a close watch on your flock when introducing new members, and deal with any problems as soon as they present themselves.

Disease

Though many flocks will likely not encounter any of these illnesses, the following diseases are witnessed more often than others.

Aspergillosis

Also known as "brooder pneumonia," aspergillosis is caused by the funguslike organism *Aspergillus fumigatus*. This organism grows easily on a number of substances and surfaces, including bedding material, feed, and rotting wood. While most healthy chickens can withstand repeated low-dose exposure, inhalation of a large volume of the spores can cause infection, as can exposure in those chickens with compromised immunity. In young

Calling the Shots

Deciding to vaccinate or administer antibiotic feed or medications should not be done without careful consideration. Overuse of any antibiotic can cause drug-resistant strains of bacteria. Vaccinating for diseases that pose no perceivable risk to your flock may prevent them from developing natural immunity to threats within their environment. While many large-scale poultry operations employ the use of both antibiotic feed and a litany of vaccinations, they often do so because they are unable to keep housing conditions as sanitary as the small flock owner will be able to. Talk to your veterinarian for information about what poultry diseases pose risk in your area, and vaccinate accordingly. Stay on top of coop cleanliness, and antibiotics may be unnecessary.

birds with acute infection, symptoms include gasping, fatigue, reduced appetite, and, occasionally, convulsions or even death. High mortality is associated more with acute cases in younger birds. Older birds seem to fare better, with the same symptoms but greatly reduced mortality. There is no treatment, so prevention is key. Remove any moldy bedding, feed, or other decaying matter. Clean feeders and waterers routinely.

Avian Influenza

The Influenza A virus subtype H5N1, also known as "bird flu," is a type of highly contagious virus occurring mostly in birds. Wild birds carry the virus naturally in their intestines and are generally unaffected by them. However, in domesticated birds, the viruses can prove especially virulent and even fatal. The virus is shed through the bodily fluids of birds, including saliva, nasal secretions, and feces. Birds then spread the virus between themselves when exposed to their secretions or excretions.

H5N1 infection manifests in two ways. The "low pathogenic" form usually produces mild symptoms, including ruffled feathers, possibly a drop in egg production, and is likely to go undetected. Conversely, the "highly pathogenic" form causes much more damage, rapidly sweeping through flocks, producing symptoms that ravage multiple organs, and incurring about 100 percent mortality within 48 hours. The virus does not usually affect humans, although a few cases have been reported. Most instances of interspecies spread of the virus occurred in individuals having close contact with birds infected with H5N1 or surfaces contaminated by the virus. As of this writing, most instances of the virus affecting both birds and humans have occurred in Southeast Asia. A vaccine has recently been approved for the prevention of human infection by one strain of the H5N1 virus. Should you have any suspicions about avian flu presenting itself in your flock, you must contact your nearest animal control agency immediately. They will assist you in determining what steps to take next.

Bumblefoot

Appearing as a large, bulbous growth on the underside of a chicken's foot, bumblefoot is caused by a cut or abrasion on the footpad becoming infected. Consider this malady to be the poultry version of getting a splinter frustratingly lodged in your foot. Once injured, whether a large cut or a tiny abrasion, the footpad begins to swell within a matter of days. It may also become reddened and hot to the touch.

The opportunistic bacteria that set in could be any one of the following: *Staphylococcus aureus, E. coli, Corynebacterium spp.*, and *Pseudomonas spp.* These bacteria are also aggressive in humans, so take caution when treating a chicken with bumblefoot. Wear disposable latex gloves, and wash clothing after contact.

It is possible to prevent your chickens from ever getting bumblefoot by keeping their feet safe from abrasion. Be certain all roosts and perches are smoothed and free of splinters. Additionally, check your coop frequently and remove any broken glass, nails, rough metal edges, or anything that could injure a chicken's feet. Heavier breeds are especially susceptible, as their weight puts extra strain on footpads. Couple that with a rough roost, and bumblefoot may not be far off.

Key to treating the illness is catching it early. If the foot is visibly swollen but is soft, wash the foot and leg, drain the abscess, rinse with hydrogen peroxide, apply an antibiotic ointment, and wrap in a bandage, changing the dressing daily. If the injury goes unnoticed and the footpad becomes hard, surgery is the only option.

Campylobacteriosis

This infection is caused by the Campylobacter bacterium. One of the most common bacterial infections affecting humans, often as a foodborne illness, campylobacteriosis can be treated with antibiotics once contracted. In chickens, symptoms include shriveled or shrunken combs, bloody or mucousy diarrhea, and even sudden death. It is possible, though, for no symptoms to be present other than a drop in egg production.

Broiler chickens contract the bacteria in the external environment, bringing it into their housing via insects, untreated drinking water, livestock and farm equipment, or on workers' clothes or boots. Once introduced to the remainder of the flock, Campylobacter spreads quickly. If it causes no symptoms, it is possible to go undetected, remaining in the birds' flesh after processing. The poultry industry is trying its best to control exposure on their end, but encourages consumers to protect themselves from the bacteria by thoroughly cooking all poultry and practicing scrupulous household hygiene when handling raw poultry.

Chronic Respiratory Disease

Also known as mycoplasmosis, chronic respiratory disease (CRD) is caused by bacterial organisms in the genus Mycoplasma. While not usually fatal, chickens that recover will remain infected for life, and future stresses may produce a recurrence of the disease. Symptoms include coughing, nasal discharge, reduced egg production, diminished appetite, sneezing, slow growth, and potentially reduced rates of hatchability and chick longevity. Outbreaks of the bacteria typically occur at times of stress when the flock's ability to resist infection may be compromised, such as when they are moved, vaccinated, too cold, or exposed to poor ventilation, ammonia buildup, or moist litter. Treatment is administered through antibiotics.

Fowl Pox

This slow-spreading viral disease can affect members of your flock at any age and at any time. Fowl pox is evidenced by wartlike nodules on the skin and in the mouth—the poultry version of chicken pox, if you will. The virus does not usually cause death, unless it impairs respiration by growing in the respiratory system itself. Since it is slow growing in nature, the virus may be present for months before symptoms appear. Once present, it affects birds for about three to five weeks. Often transmitted initially by mosquitoes, the virus spreads within the flock by both direct and indirect contact with infected birds. Chickens that recover do not remain carriers of the virus. As routine sanitation and hygiene practices will not prevent it, vaccination is often performed as a proactive measure. There is no treatment for fowl pox.

Fowl Typhoid

This is a highly infectious, contagious bacterial disease caused by the bacteria *Salmonella gallinarum*. Mortality is often 100 percent. Although occurring most often in young birds, the disease can affect chickens of all ages. Signs of infection include lethargy, reduced appetite, increased thirst, pale combs and wattles, green or yellow diarrhea, or sudden death. Fowl typhoid is spread by droppings from infected birds, as well as by contaminated food, water, clothing, and equipment. Antibiotic treatment exists, although it will not remove the infection. A vaccine may be preventatively administered.

Infectious Bronchitis

Caused by a virus affecting only chickens, infectious bronchitis is a highly contagious respiratory disease. When it appears, any and all susceptible birds around will become infected. The disease spreads through airborne transmission, and can also be carried on clothing, equipment, and metal surfaces. Symptoms include coughing, gasping, rattling, sneezing, nasal discharge, and severely reduced egg production. If it affects young birds, it may damage reproductive organs, rendering them unable to produce normal eggs. An outbreak affecting a laying flock halts egg production to almost zero. Those eggs that are laid may be small, soft-shelled, and oddly shaped. As there is no treatment, prevention through vaccination is the only means of keeping the illness at bay.

Infectious Coryza

A respiratory disease caused by the bacteria *Hemophilus gallinarum*, infectious coryza typically occurs in semi-mature or

adult birds. It may present as slow growing in nature, affecting only a few birds at a time, or spread quickly, infecting a greater number of chickens at once. Telltale characteristics of the disease include swelling of the face and wattles, smelly nasal discharge, watery discharge from the eyes (which may cause the eyelids to stick together), a drop in egg production, impaired vision, and reduced food and water consumption. The disease is usually introduced via carrier birds added to healthy flocks. Once a chicken is infected, it remains a carrier and shedder of the bacteria for life. Infectious coryza may be treated with antibiotics. For disease prevention, an all-in, all-out policy is advised, which basically means culling existing birds before adding to your flock. Otherwise, either breed your own chicks or purchase new birds and chicks only from highly reputable hatcheries and sellers.

Infectious Laryngotracheitis

Infectious laryngotracheitis (ILT) kills cells lining the airways of a chicken's windpipe, resulting in varying degrees of impaired respiration. ILT is caused by a herpes virus, and birds that recover may become lifelong carriers, often transmitting the disease in flocks. While ILT does not affect humans, they can spread the disease to chickens via clothing, dirty footwear, or poultry equipment. Symptoms include coughing, gasping, discharge from eyes and nostrils, or even death. The disease is highly contagious, and mortality can be up to 70 percent. There is no treatment, so prevention is essential. Only buy birds from a reputable hatchery whose birds are known to be ILT-free. If adding birds to your flock, quarantine them first for up to a week to see if any disease symptoms manifest.

Marek's Disease

Caused by a virus in the herpes family, Marek's disease usually affects young birds. It is believed to live in the feather follicles and shed in dander, or sloughed-off skin and feather cells. Spread of the infection is usually respiratory, and the rate of contagion is high. Marek's disease may manifest in a number of ways. Neurologically, it results in various forms of paralysis. Viscerally, tumors infiltrate numerous organs, including the heart, reproductive organs, muscles, and lungs. If the disease presents itself on the skin, tumors of the feather follicles may result. In acute cases, the disease may advance very quickly, killing birds previously exhibiting vibrant health. Most hatcheries vaccinate one-day old chicks against Marek's disease.

Newcastle Disease

Newcastle disease is a widespread, highly contagious viral respiratory disease affecting all species of poultry. One strain of the virus, referred to as "exotic Newcastle disease" is considered such a threat that strict border control measures exist to prevent its entry into various countries around the world. Once exposed, almost all birds in a flock will become infected in a number of days. Symptoms are similar to other respiratory diseases in poultry and include coughing, sneezing, gasping, difficulty breathing, and, if allowed to advance, paralysis. Mortality rates vary widely, from zero to 100 percent. No treatment exists for Newcastle disease. Vaccination is widely practiced by large-scale poultry producers. The disease can survive for long periods in normal temperatures, so be sure to thoroughly clean coops with disinfectant and hot water if you have an infected chicken.

Parasites

Various parasites, both internal and external, are known to infect chickens.

INTERNAL

Coccidiosis

Coccidia are protozoan parasites that live in the cells lining the intestines of all poultry. In most instances, it lives in a symbiotic relationship with chickens. However, under unsanitary conditions, it can proliferate extremely quickly, resulting in potentially grave danger to birds. As it most commonly affects chicks, refer to chapter 7, Raising Chicks, for more information.

Worms

Transmitted by wild birds, rodents, and insects, worms are fairly common in chickens, and don't always pose a problem. However, if one or more of your birds exhibit diarrhea, listlessness, weight loss, and a drop in egg production, you might have a worm situation that has gotten out of hand. Roundworms, tapeworms, and cecal worms are the most common types of worms associated with chickens. If you suspect worms, don't worm indiscriminately. Doing so can result in the development of drug-resistant colonies of parasites, as well as keeping your chickens from forming their own resistance to parasites. Bring a sample of droppings in to your veterinarian for testing first.

EXTERNAL

Lice

In chickens, lice can be found either on the head or on the body. They are visible to the naked eye and resemble small, tiny to almost transparent organisms. When the chicken's feathers are parted, you will see lice scurrying about if your bird is infested. Symptoms indicating you may have a lice problem include: witnessing one or more of your chickens trying to pull out their feathers (to stop the itching and irritation), a reduction in egg

It is important to check for external parasites on a regular basis.

production, weight loss, and a dirty backside. If one bird exhibits symptoms, then treat your entire flock, as they will likely all have lice. These nasty critters are more often an issue for birds permitted to free-range or having access to the outdoors than for birds kept intensively. Encourage your birds to dust bathe, as dust clogs louse pores. Treatments include dousing your flock with louse powder or a chemical spray. Check with your local agricultural extension office or veterinarian for an approved poultry insecticide. Other natural options I have encountered include liberally dusting with sulfur, diatomaceous earth, and wood ashes.

Mites (Red, Northern Fowl, and Scaly Leg)

Mites are parasites that live either in the blood of chickens, or burrow deep into their skin or in their feathers. Once they infest a chicken, they cause delayed growth, a drop in egg production, irritation, blood loss, diminished fertility, and on occasion, death.

Dusting Themselves Off

Dust baths are essential for maintaining health and wellness in your flock. A dust bath is an area of dry, sandy soil. Your chickens will either create this area themselves, in the warmest, sunniest portion of their coop, or you can provide it for them. If space is limited in your yard, you'll want to keep their dust baths confined to a certain area (and out of your veggie beds!); or if your birds are intensively kept, simply fill up a deep tray, wooden box, or even a shallow terracotta planter with loose dirt and dry sand. Dust bathing clogs the breathing pores of parasites that prey on chickens, serving as the most perfectly natural and organic insecticide imaginable. Accordingly, this habit should be encouraged. Don't worry, dust baths don't result in dirt-crusted chickens. To the contrary, after digging themselves as far down into the dirt as possible, chickens puff out their feathers, shake themselves off, then begin preening.

Red mites, found mostly in warmer weather, come out at night. They live in henhouse crevices and can be seen crawling over walls, roosts, and chicken's bodies with the help of a flashlight. Red mites are so named on account of the way in which their normally grayish-white bodies turn red when engorged with blood. Treat them by cleaning your coop thoroughly and liberally dusting with an approved mite insecticide.

Northern fowl mites appear during cold weather and crawl on birds during the day. Discourage these mites by placing cedar chips, a natural mite deterrent, in nesting boxes as bedding.

Scaly leg mites invade the legs and feet of chickens, burrowing in and causing scales to protrude and fall out. An affected chicken, if left untreated, may eventually have difficulty walking. One treatment involves rubbing the legs with an equal mixture of kerosene and linseed oil. If the thought of rubbing kerosene on a bird and the subsequent fire hazard that doing so presents scares you off this treatment, alternatives include rubbing petroleum jelly or some other lubricant over the infected bird's shanks weekly for up to two weeks, or dipping their shanks weekly in some form of vegetable oil. Telltale signs of the presence of mites other than actually viewing them include "egg spotting," which are specks of blood on egg shells where mites were squashed by sitting hens, and "salt and pepper" deposits underneath perches.

Cannibalism

As alarming as it sounds, several forms of cannibalism are common among chickens. Naturally, you will want to stop these behaviors in their tracks for a healthy, unmolested flock.

Egg Eating

Egg eating is a form of cannibalism that starts when a hen accidentally breaks open an egg, samples the goods, and realizes she likes the taste. After that, others may get in on the action and begin breaking open eggs deliberately. Egg eating may result from too crowded living quarters or from eggs left in nesting boxes for too long. Discourage this nasty habit from ever beginning by emptying nesting boxes early and relocating the instigator (the one with the yellow goop festooning her beak and the "who, me?" look across her guilty face) to a new home. You can discourage future egg eating by placing bogus "eggs" that have been blown empty and filled with mustard inside the nesting boxes.

Feather and Toe Picking

Chickens pick and peck, 'tis their nature. Which is normally all good and well, but when the picking and pecking occurs on each other and not on grains of corn or bits of a wriggling grub, an otherwise natural habit can quickly become lethal. If you notice one of your birds missing feathers, attempt to determine if she is doing it to herself. If she is, try to find out what is giving her the itchies. If you have a rooster, the loss in feathers could be the result of a particularly passionate leading man. Also, consider that your bird may be molting. Ruling out roosters, critters, and molt, the cause of feather loss might be a hen's flock mates.

A chick's toes can look a good bit like little yellow worms. A curious chick might pick at his or her own toes or those of their flock mates and, at the sight of blood, go on a pecking free-for-all. Older chickens are more inclined to pick the heads and tail areas of other birds. Either way, once there is blood, curious picking can spiral into cannibalism. Picking can be caused by cramped living conditions, a nutritional deficiency, lack of sufficient feeders and waterers, harsh lighting, too much heat in a brooder, external parasites, inadequate ventilation, or some other stressing vari-

able such as inclement weather, improper handling, or moving to new living quarters. A cannibalistic bird might also simply be a bored bird. While this is not usually the case in birds permitted access to the outdoors, those kept indoors or in cages for extended periods of time denied the opportunity to express their natural inclination to scratch, dig, hunt, forage, and otherwise pick will manifest this urge in less desirable ways.

If you discover cannibalism within your flock, curb it immediately. Attempt to discover the root of the problem. Consider giving your birds distractions by tying pieces of fruit or vegetables to strings and suspending them from the coop ceiling or offering scratch grains for them to turn their picking attention toward. Always sequester injured birds as discussed early in this chapter. Try rubbing vinegar or a poultry-specific ointment on the feathers and wound of the chicken on the receiving end of the bullying; flock mates will get a mouth full of nastiness when they take an opportunistic peck. If all else fails, you may have to resort to either removing the instigator from your premises or permanently rehousing the injured bird in separate quarters.

Other Ailments and Conditions

Although this list is not comprehensive, there are several other things to watch out for in terms of your flock's health and wellness.

Acute Death Syndrome (Flip-over Disease)

Have you ever heard about someone who "up and died"? Chickens who meet their end this way are said to have flip-over disease. Occurring most often in intensively confined male birds bound for table, the illness causes sudden death, prefaced by a brief wing-flapping convulsion. Although the cause is unknown, it is believed to be metabolic in origin, most likely the result of overfeeding. Many broilers overeat, growing rapidly all the while, causing sudden heart failure from ventricular fibrillation. No warning signs foretell an impending demise. Otherwise healthy chickens, sometimes in

midstep, literally flip over and die on their backs. Incidents of occurrence may be reduced by changing the diet of broilers to mash (thereby lowering carbohydrate consumption), restricting feed availability, introducing artificial light that mimics daylight and evening hours, and reducing stress.

Crop-binding

It is possible for a chicken's crop to become obstructed, or blocked, preventing food from moving out of the crop and into the proventriculus. When this occurs it is usually the result of consuming long blades of grass, bedding material, or feathers. The blockage will be visible as a large bulge in the center of a chicken's breast, but only in the evening, after the chicken has had a full day to eat and fill up their crop with other material. Left unattended to, the affected bird's breathing may become difficult, potentially resulting in death.

If you notice a large bulge in the evening that is still there the following morning and is accompanied by a lack of appetite,

try giving your chicken either a little warm water or vegetable oil and then gently massaging the surface of the bulge. Use caution when attempting to put any liquid down a chicken's throat, as they can asphyxiate if given too much liquid too quickly. It the lump does not loosen and break up, a visit with the veterinarian may be necessary.

Egg Binding

If a chicken attempts to pass an egg that is especially large, it can become lodged inside her vent. Also, chickens receiving inadequate exercise can be susceptible to becoming egg bound, as lack of activity may cause the buildup of a fatty layer around her reproductive organs. This in turn impedes the ability of the oviduct to push eggs down toward the vent. Should you witness one of your birds squatting with nothing coming out, hanging out in the nest all day, or looking as though she may be, well, constipated, it's possible she may be egg bound. Cover your finger with mineral oil, warm olive oil, or some other lubricating substance safe for human skin, and insert it into the vent. If you can feel an egg inside, rub the lubricating substance all around the vent, which should assist her in passing the egg.

Attempt to get the egg out of your chicken, carefully massaging her abdomen in the direction of the vent. You didn't know you signed up for this when you took on chicken raising, did you? As a measure of last resort, you may need to puncture the egg while it is still inside the chicken, carefully removing it in small pieces. This carries the risk of injuring her vent, though, and should only be done if nothing else seems to get the egg to move. Clean the vent afterwards with hydrogen peroxide in a spray bottle to prevent possible infection.

Molting

You know how snakes shed their skins every year and some dogs shed their winter undercoats come spring? Well, chickens also have an annual shedding period, called molting. During this process chickens gradually lose their feathers. Prompted by waning daylight hours, molting takes place in most chickens around late summer to early fall, although some molt in early winter. As their feathers fall out, new feathers called "pinfeathers" come in, resembling short, hollow drinking straws at first. Molting follows a sequence, beginning at the head, then moving down the neck, to the body, on to the wings, and finishing at the tail. In most chickens, the process takes about six weeks; however, it is not unheard of for some to take up to 12 weeks to molt. While some chickens molt so slowly you may not even be aware the process is occurring, others, perhaps the more scandalous of your flock, toss off all their feathers at once, running around partially nude with abandon. Egg production customarily drops off during this time and can even halt completely in some chickens. Molting is stressful, so don't be surprised if an otherwise happy-go-lucky hen suddenly turns a bit skittish and cross; the attitude generally dissipates with the arrival of the new plumage.

Portrait of a chicken owner

Kevin

When he's not busy serving as director of archaeology at the home of a former U.S. president, Kevin and his wife oversee a flock of five hens and one rooster from their suburban Tennessee home. His interest in keeping chickens stems from a two-fold desire to both provide quality protein for his young son as well as reduce the distance his food travels from farm to table. "We source the majority of our food within 50 miles of our home. And, if we have difficulty finding products, we try to grow, raise, or produce them ourselves." Large-scale poultry operations are notorious polluters. Kevin and his family are able to opt out of participation in that industry by housing chickens on their property.

Keeping chickens has enabled Kevin to gain a bird's-eye view, if you will, of ecosystem cycles. By offering table scraps to his flock, which in turn reciprocally provide eggs and droppings to be used for compost in his garden beds, Kevin has seen "the direct connection between what we feed our hens and what they provide for us." Furthermore, by keeping a small flock, he hopes to impart lessons in balance, sustainability, and stewardship to the next generation, starting right at home with his son. Kevin hopes that living close to his son's food source will engender in him a greater respect for and appreciation of the natural world and the position of humans within it. While he is aware of the idealism inherent in his aspirations, he stands by them, steadfastly and earnestly asserting that in so doing, he and his wife "truly believe that we are forging a better future."

Chapter 9
Eggs

Whether served up tableside, hidden outdoors for Easter, or artfully crafted into decoration, eggs are an integral part of cultures the world over. From folklore to holiday games, people have woven eggs into the fabric of their lives. Nutritious, versatile, and awe-inspiring, it's easy to see why we hold eggs in such high esteem.

WHICH CAME FIRST?

Female chickens are born with two ovaries. When she is still young, the right ovary stops maturing, leaving the left to carry on the work involved in egg formation. This left ovary contains 4,000 or so ova, which are undeveloped yolks. As a pullet reaches the point where she is ready to begin laying eggs, these ova begin to mature, one after the other. Ova are mature once they have acquired sufficient layers of yolk. They then slip into the oviduct, a long tube terminating in the vent, which is the same exit point for droppings. This process is called ovulation.

As it travels along the oviduct, a number of things happen to the yolk. If a hen has recently found herself the recipient of a rooster's passionate embrace, sperm may be present and the yolk will become fertilized. Otherwise, the yolk becomes shrouded in egg white, technically referred to as albumen, gets wrapped up in fibers called chalazae that anchor the yolk within the white, and lastly is enclosed in a pigmented shell.

The turnaround time of egg production varies slightly from hen to hen, averaging 25 hours, with some hens laying closer to every 24 hours and others preferring 26 hours. My Ladies all lay at different times of the day. Since each one proudly announces her success with barely contained enthusiasm, I am inundated with joyful proclamations all day long. Hens prefer not to lay in the evening, causing them to occasionally miss a day and lay the following morning. How many eggs a chicken will lay depend on a number of factors, including breed, age, climate, and stress level. Chickens intended to be heavy layers are bred to have the shortest turnaround time between eggs. Such chickens can produce 300 eggs a year when they are at peak production age. Purebred chickens will more often lay approximately 250 eggs annually. Production will wane as daylight decreases in the winter months.

SIZE, SHAPE, AND COLOR

Size and shape of eggs is informed by a chicken's age. A pullet's first eggs may be tiny, somewhere between the size of a marble and a grape. As she ages, the size of her eggs will increase, eventually weighing 2 ounces (56 g) on average.

As for shape, irregularities during formation can result in some odd-looking eggs, including those with wrinkles, exaggeratedly pointed ends, or thin shells. Eggs may also gain odd appearances as a chicken ages or if she becomes scared. Don't be distressed if you find the occasional goofball eggshell, especially if you are dealing with pullets or seasonal weather variations. These are fine to eat, but don't hatch them, as you don't want to pass on that trait.

The color of a chicken's eggs corresponds to the color of her ear lobes, which are found just behind the eyes on each side of the head. As a rule of thumb, hens with white ear lobes will lay white eggs, while those with red ear lobes will lay brown eggs, although some variations to this exist. Generally speaking, Mediterranean-originating breeds will lay white eggs, while Asiatic and most American breeds will lay brown eggs. In the middle, you find a spectrum ranging from almost pink to blue-green.

UNUSUAL EGGS

Occasionally, you will come across an abnormality in an egg. Most of the time, it will be a one-time occurrence, requiring no further action on your part other than monitoring your eggs to see if the anomaly returns. Sometimes, though, weird eggs can be a sign of distress in your birds. When a pullet first begins laying eggs, she may produce an egg missing a yolk. These are called no-yolkers, dwarf eggs, or wind eggs. This rarely happens in mature hens, but it certainly isn't outside the realm of possibility. Pullets can also produce eggs with two yolks, aptly known as double-yolkers, at the beginning of her laying cycle. Don't be concerned about no-yolkers or double yolkers. Your pullets are merely setting up their production cycles. If this continues, though, you may consider a visit with your veterinarian to rule out potential problems.

Several other unusual features can show up inside the eggs from time to time. Blood spots or meat spots, while they may look unappetizing, do not compromise the integrity of an egg. Both are remnants of reproductive tissue that broke free from a blood vessel during egg formation. They may be indications that your hen is receiving inadequate vitamin A in her diet. A tendency to produce eggs with blood spots can be hereditary,

Salmonella

Salmonella enteriditis is a type of bacteria present in the feces of many animals, including chickens, that is known to cause severe food poisoning in humans. The bacteria can be transmitted to eggs either through the porous shell or inside of the egg itself, perhaps when eggs are being formed inside a chicken's ovaries. Salmonella poses a risk if ingested raw, so cooking eggs to 160°F (71°C) is considered essential, especially if your eggs are purchased from a large-scale egg producer. For the most part, though, the likelihood of contracting salmonella through a small, home-based operation is rare. Staying on top of coop cleanliness drastically minimizes the risk of exposure. Prompt removal of eggs from their nest boxes, followed by refrigeration, also lessens the chances of salmonella contamination, as the bacteria multiplies at room temperature. Be sure to wash any utensils that have come into contact with raw eggs thoroughly after use, and always wash your hands with hot, soapy water when handling eggs.

so depending on whether your eggs are for your own personal use or for sale to the public, you may opt to cull the responsible party. Nature isn't always fair, I know.

If you come across funky-tasting eggs, consider what table scraps you have given your flock recently. Flavors from garlic, onions, and fish can be imparted to eggs, resulting in an off flavor. As eggs are porous, they can also pick up strong odors from any chemicals near the chicken coop, such as kerosene, gasoline, mold, or dank scents. Lastly, if you come across (so sorry for having to write this) worms in an egg, take immediate action. You're dealing with parasites here and need to get the afflicted bird on the mend pronto and do all you can to curtail

the same fate befalling the remainder of your flock. Call your vet, and disinfect your coop.

QUALITY CONTROL

If you have free-ranging hens, you may find eggs in assorted and sundry locations whose freshness is questionable. There are a few tools available for determining whether the egg is fit to consume. Using a candling light, you can do one of two things. First, you can check yolk visibility by spinning the egg in front of the light. A fuzzy-looking yolk means the egg is fresh, whereas a clearly defined yolk indicates aging. Also, you can look for air-cell size. A fresh egg has no air cell, while the size of an air cell increases as the egg contents cool and shrink. Refer to page 80 for more information about candling.

Other means for determining freshness include floating, smelling, and simply examining the contents by cracking the egg open. Floating is exactly what it sounds like. A fresh egg will sink to the bottom of a bowl of water and lie there horizontally. An egg that is around one week old, thereby containing a growing airspace, will rise up on a slight diagonal. Eggs two to three weeks old will lift vertically, with the tip resting against the bottom of the bowl. Older eggs will float right up to the top on account of their large airspace. You can also simply sniff for any rotten-egg odors, which indicates the presence of hydrogen sulfide and an egg that is long past its prime. Finally, cracking open an egg will give you ample information about its age. In general, fresh eggs have firm yolks and cloudy whites while older eggs have yolks that are more likely to break and watery, runny whites.

WELL-ROUNDED MEALS

Eggs are absolute nutrient powerhouses. They are often referred to as the perfect food, and for good reason. Although a large egg contains only about 75 calories, it is laden with all eight essential amino acids; vitamins A, B12, D, and E; folic acid; phosphorus; and zinc. One egg contains roughly 15 percent of the U.S. recommended daily allowance of protein. This protein is widely considered to be of superior quality among food proteins, second only to that found in human breast milk. Egg yolks are composed of

Fresh eggs are full of nutrients.

fats, cholesterol, and pigment, along with several other nutrients. One of the fats, lecithin, purportedly plays a crucial role in brain function. Eggs also contain the nutrient choline, which is said to play a pivotal role in fetal brain development and the prevention of birth defects. The carotenoids lutein and zeaxanthin are present in eggs, which research suggests assist in eye function and integrity.

There are some people who refrain from eating eggs, believing them to cause elevated blood cholesterol levels and subsequently increasing the risk of heart disease. Cholesterol is a fat required by humans for a number of functions, including regulating hormones, assisting in brain function, and converting sunlight into vitamin D. Most cholesterol present in humans is produced internally, with only about 20 percent coming from dietary sources. A healthy body regulates dietary cholesterol, eliminating or adding it as needed. Numerous studies have shown that eating eggs has no effect on blood cholesterol. A growing number of scientists believe trans fats and saturated fats are the more likely culprits in raising blood cholesterol levels than are dietary sources of cholesterol. Furthermore, the choline present in eggs works to break down the amino acid homocysteine, which research has indicated may contribute to an increased risk of heart disease.

Sharing and Selling

Depending on the size of your flock, it's possible you may have more eggs than you can reasonably scramble, poach, or curd. Selling your eggs to coworkers or neighbors is a time-honored way to unload your Ladies' bounty. Better yet, consider trading or bartering with those neighbors or coworkers. Everyone has a skill or service they can offer, even if it's something as simple as exchanging your eggs for a loaf of someone's homemade bread. Ask around and see what folks might be willing to trade.

Perhaps you have enough surplus eggs that you are thinking of selling them at a local farmer's market or corner grocer. If you intend to market them in any particular way, such as "free-range" or "certified organic," be aware that laws exist related to such labeling claims. Contact your county Extension office to learn what the regulations are in your area before you start proffering your eggs on a large scale.

STORAGE

Gather up your flock's eggs at least once a day, if not twice (or more!). Doing so not only prevents egg eating, but also keeps the eggs from getting sullied and ensures quality. Eggs that are soiled can be wiped off with a dry cloth, rough paper towel, or a bit of superfine sandpaper. It's best not to get eggshells wet, as water will rinse off the protective covering, or "bloom." Truly soiled eggs are better disposed of than rinsed. If you keep your nesting boxes clean, this really shouldn't be an issue anyway. As a stay-at-home writer, I have the luxury of checking on my Ladies three times a day. I realize that most people aren't home all day, though, so gathering eggs first thing in the morning is your best bet.

Stored in a cool, dark place at about 70 to 75 percent humidity, eggs will keep for up to five to six weeks. If you store your eggs in the refrigerator, place them on the lowest shelf, as this is generally the coolest place. Most refrigerators have low humidity, causing foods to dry out, so eggs kept in a fridge will last for about four weeks. Always store eggs in a closed container with the pointed end facing down to keep the yolk centered. If you have a surplus of eggs that you know you won't use in time, crack the eggs into a freezer-appropriate container and store frozen for up to one year. Alternatively, you can separate yolks and whites and freeze in ice cube trays, allowing you to thaw out only what you need. Either way, sprinkle a pinch of sugar or salt to keep the eggs from getting gummy when they thaw. Never freeze eggs in their shells, or they are likely to explode. I don't know about you, but I can think of, oh, a thousand things I'd rather be doing than removing frozen eggshell bits from the interior of my freezer.

A Good Way to Dye

Naturally dyed eggs make a colorful addition to any festive occasion, whether you celebrate Easter, Beltane, or Passover. Eggs can be dyed using many household fruits, vegetables, herbs, and spices, creating subtle shades that are nonetheless striking in their simplicity. To create a dye bath, add the dying agent of your choice to a large stainless-steel pot filled with 1 quart (1 L) cold water and 2 tablespoons (30 mL) white vinegar. Bring to a boil, then reduce heat and simmer for 25 minutes. Next, add however many raw eggs you would like to dye to the bath, and boil at least 30 minutes. Remove eggs from bath and dry gently with a paper towel or old cloth. Need ideas? Here's a color chart to get you started:

Color	Agent
Pale yellow	Orange marigold leaves
Golden yellow	Turmeric
Pale purple	Cranberries
Blue-purple	Blueberries
Lavender	Raspberries
Pale pink	Beets
Dark pink	Red cabbage
Dark brown	Coffee grounds
Copper	Onion skins

A World of Eggs

Creation stories the world over place eggs at the center of their mythology. In the traditional Chinese creation story, the first living thing, called Pan-gu, grew inside a gigantic cosmic egg. All elements of the universe, including male and female, wet and dry, heaven and earth, and light and dark, originally co-mingled inside the egg with Pan-gu. After 18,000 years, the egg hatched, and the universe formed into distinct components. Similar mythology underlies an Egyptian creation tale. In this story, the Chaos Goose and the Chaos Gander create an egg that develops into the sun, or Ra. Along with the Egyptians, ancient Persians, Phoenicians, and Hindus also believed the world to have originated initially from one egg that split and created the world as we know it today.

Alchemy, the medieval practice of chemistry and philosophy, revered the egg as well. When attempting to craft the "philosopher's stone," a substance believed capable of turning all other metals into silver or gold, alchemists placed their materials in an egg-shaped vessel titled the "philosopher's egg." It was believed the shell, membrane, yolk, and white mirrored the four natural elements of earth, air, fire, and water, components the alchemists believed to be essential for the successful transmutation of materials.

Emblematic of life and fertility, eggs were exchanged as gifts by ancient Romans, pagans, and Egyptians. Their symbolic value continues in the tradition of hunting for and eating eggs—be they chocolate, colored, or otherwise—for Easter. Christians incorporated the egg into Easter celebrations to symbolize the unification of all God's creatures. In Czarist Russia, eggs were given as gifts throughout the year. Similar in design to well-known *matrioska* dolls, wooden eggs-within-an-egg, called "nest eggs," were given to mark special occasions such as 18th or 21st birthdays, weddings, or other significant events. Perhaps the most famous Russian egg-related legacy is the jewel-encrusted eggs created by Carl Fabergé. The son of a Swiss immigrant to Russia, Fabergé served as the imperial family's jeweler at the turn of the 20th century. With the aid of his talented crew, he crafted ornate eggs the family bestowed upon one another for special occasions.

Portrait of a chicken owner

Lynne and Bruce

Folks flock to Lynne and Bruce's eggs. From their 5½ acres (2.23 hectares) in southern California, the couple keep a watchful eye over their 100-plus chickens, whose eggs supply a nearby restaurant with over 20 dozen eggs weekly. Jay Porter, owner of The Linkery restaurant, had been searching for a local supply of pastured eggs. The restaurant places a great deal of emphasis on procuring their items locally, serving only meat, vegetables, fruits, and eggs from nearby farms. Lynne explains that Jay **"not only wanted to support local farmers but also wanted to educate his customers as to where their food originated."**

Initially, Lynne and Bruce furnished the restaurant only with eggs. They have since extended their offerings, now growing fruits, vegetables, and herbs exclusively for themselves and the restaurant. Not only is their operation with The Linkery a mutually beneficial financial endeavor, it is one that is environmentally beneficial as well. In addition to helping conserve nonrenewable resources by offering locally produced food, Lynne and Bruce live and operate their business in a sustainable manner. Droppings from their chickens, as well as from goats they keep for making their own cheese and yogurt, are used to fertilize their vegetable beds. The water used to hydrate the crops is from a well powered by a solar pump. According to Lynne, **"we are self-sufficient and self-sustaining."** Their main motivation for keeping chickens, however, might just be because of the way in which chickens seem to serve as a pressure valve release for so many people. **"Watching chickens daily is better than a therapist,"** Lynne enthuses.

Chapter 10
Recipes

Eggs! Whether they're savored along-side morning coffee and jam-smeared toast, rendered into a delectable frittata for lunch, or transformed into succulent deviled bites for an anytime snack, eggs are truly one of our most versatile foods. Old standbys rub shoulders with modern variations in this chapter, all of them guaranteed to become your new go-to recipes. Sweet, herbaceous, fluffy, boiled, or scrambled, you'll find a little some-thing here for every cook.

The Perfect Boiled Egg

The boiled egg just might be the unsung hero of the
cooked egg repertoire. Imminently transportable, easy
to cook, and capable of numerous applications, hard-
and soft-boiled eggs offer broad appeal. Follow these
steps to ensure that your eggs are cooked properly,
and then customize them to your preferences. For me,
that would be chopped up and mixed into the most
delicious egg salad ever, or simply cracked open and
enjoyed in a decorative egg cup. Delicious!

Eggs (any number, just be certain they are at least one week old; see tip, below)

A stainless steel pot with a lid

If making hard-boiled eggs, you will also need:

Mixing bowl

Ice

Cold water

TO BOIL EGGS:

→ 1. Fill the pot with water and gently place the eggs inside. Cover with the lid. Bring to a roiling boil over high heat. Allow to boil for 2 minutes, then turn off heat. Leave the pot covered on the stovetop for 2 minutes for soft-boiled, 15 to 20 minutes for hard-boiled.

→ 2. Serve soft-boiled eggs immediately.

→ 3. For hard-boiled, while the eggs rest, fill the mixing bowl with cold water and ice. Remove the eggs from the stovetop and transfer to the ice-water bath. Leave in the water for 4 to 5 minutes, then peel. If you do not plan to eat your eggs at this time, store in the refrigerator, unpeeled, for up to one week.

→ 4. To serve, crack hard-boiled eggs on the side of a bowl and peel off the shell. It may be necessary to rinse peeled eggs under cold water to remove any shell bits.

TIP:

When eggs are fresh, they generally have a low albumen pH. This low pH causes the albumen to stick to the interior shell membrane more strongly than it sticks to itself. After several days of refrigeration, either at a grocery store or at home in your own refrigerator, the pH rises to about 9.2, allowing the shells to be peeled more easily.

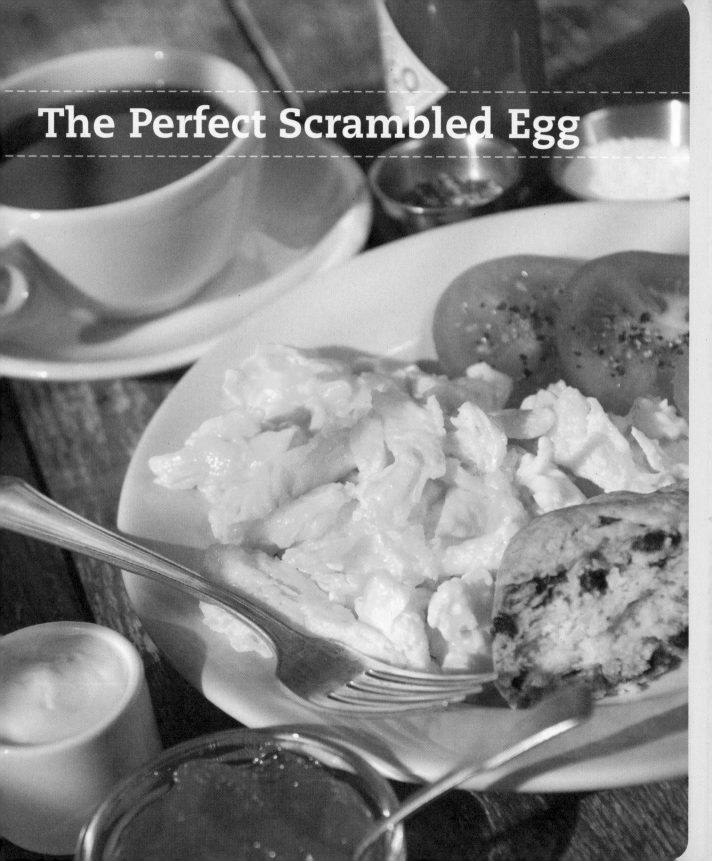

The Perfect Scrambled Egg

Although scrambling eggs seems pretty straightforward, it's actually quite easy to overcook or improperly cook them, resulting in dry, dense eggs. The key to perfectly scrambled eggs lies in beating them thoroughly before cooking. This works air into the eggs, helping to turn them into fluffy curds. It is also imperative that you have every ingredient you intend to incorporate into your eggs prepped in advance. Scrambled eggs cook up quickly and there will be precious little time to grate cheese or chop herbs along the way. Finally, be sure to remove the eggs from the heat just before they look completely cooked. Eggs that look done in the skillet are almost sure to be overdone when plated. Eaten plain, tossed with a handful of fresh chopped herbs, or drowned in ketchup, any way you serve them, scrambled eggs please every palette! *Serves 2.*

YOU WILL NEED:

- 4 eggs
- 1 tablespoon (15 g) butter
- Sea salt
- Freshly ground pepper
- Cheese, fresh or dried herbs, tomatoes (all optional)

TO PREPARE:

1. In a medium mixing bowl, beat eggs vigorously with a fork or whisk for 20 seconds.

2. Place a medium nonstick skillet over medium heat. Melt the butter in the skillet until foamy and bubbling. Add a pinch of salt to the egg mixture, beat to incorporate, and then pour into skillet.

3. Using a rubber spatula, slowly stir the eggs toward the center of the skillet, tilting the pan to move around any runny eggs. If using any optional ingredients, add them at this point. As soon as curds begin to form, stop stirring.

4. Using the spatula, begin to slowly fold the curds over on top of each other. Once there are no more liquids visible in the skillet, immediately remove the pan from the heat.

5. Grind pepper on top of the eggs and serve at once.

The Perfect Omelet

Making a perfect omelet need not be as difficult as it's often cracked up to be. It really is simply a matter of having the right pan, at the right temperature, and the right technique. Made with a limitless array of fillings, omelets can be new again each time you prepare them.

This recipe produces a thin, moist, perfectly browned omelet, which complements vegetables, cheeses, and herbs wonderfully. Follow the instructions carefully, and your days of dry, cracked, mediocre omelets will quickly become a thing of the past. *Serves 1.*

2 tablespoons (30 g) butter or olive oil

2 to 3 eggs, depending on the size of your eggs

Pinch of salt

Dash of hot sauce

Fresh herbs

Cheese, such as goat cheese, feta, cheddar, or swiss, if desired

Vegetables, such as tomatoes, olives, zucchini, mushrooms, or peppers, if desired

Cubed ham, smoked salmon, or sausage, if desired.

Feel free to combine whatever herbs are seasonally available and local or use them singularly; a favorite blend in omelets at my house is marjoram, basil, thyme, and a bit of lavender—delicious!

formed across the surface of the mixture, and it will no longer look "wet" in the skillet.

5. Put fillings and cheese on half of the mixture. If using fresh herbs, place them in now as well. Using a wide spatula, fold the unfilled side of the omelet over the side with fillings, and cover with the lid for 1 minute longer.

6. Remove the lid. Using the spatula, carefully slide the omelet out of the skillet and onto a plate, and serve immediately.

TO PREPARE:

1. If you intend to fill your omelet with vegetables, go ahead and sauté them in a bit of olive oil or butter now. If you attempt an omelet without first cooking the fillings, the result will be undercooked, bland vegetables inside. Once you are done sautéing, set the vegetables aside.

2. In a medium mixing bowl, whisk the eggs with the salt and hot sauce. Fresh eggs need to be agitated more intensely than older eggs in order to properly emulsify, so be sure to whisk until your eggs are light and frothy. If using dried herbs, whisk them in now.

3. Place the skillet over medium-low heat. Add either butter or olive oil to the pan to coat the entire bottom of the skillet. Pour the egg mixture into the skillet. Gently tilt the pan by its handle, and swirl until the eggs are distributed evenly.

4. Put the lid on the skillet for 2 minutes. Every 30 seconds or so, turn the pan clockwise 90° to ensure proper heat distribution, as some stovetops and pans heat unevenly. Remove the lid. Small bubbles should have

Freshly grated organic cheese

Herbed Deviled Eggs

Long considered summer picnic fare, deviled eggs can be served up all year long. While this recipe features tarragon, you could easily modify to make use of whatever seasonally available fresh herbs you have on hand. Try thyme in autumn, sage in winter, and parsley with chervil in spring. Whether you are shoveling snow or weeding the garden, deviled eggs are the perfect treat year-round! *Yield: 24 servings*

YOU WILL NEED:		YOU WILL NEED (continued):	
	12 hard-boiled eggs		Sea salt
	2 tablespoons (30 mL) mayonnaise		Freshly ground black pepper
	2 tablespoons (30 mL) extra-virgin olive oil		Dash of hot sauce
	2 tablespoons (30 mL) Dijon mustard		Tarragon leaves or small bits of pimento (optional, for garnish)
	2 tablespoons (30 g) sweet pickle relish		
	1 tablespoon (15 g) capers		
	1 teaspoon (5 g) lemon zest		
	1 tablespoon (15 g) fresh tarragon, chopped		

TO PREPARE:

→ 1. Remove shells from eggs; rinse under cold water. Cut eggs in half lengthwise. Carefully remove yolks and transfer to a medium bowl; set whites aside.

→ 2. Add remaining ingredients to yolks. Mix with a fork or whisk until creamy and light.

→ 3. Either fill a pastry bag and pipe filling into egg-white halves, or simply portion filling out with a spoon into each half. If using garnish, top each egg-white half with either a tarragon leaf or a piece of pimento. Serve immediately, or refrigerate for up to three hours.

Southern Deviled Eggs

For a more traditional deviled egg, omit capers, lemon zest, and tarragon. Substitute prepared yellow mustard for the Dijon mustard. Fill egg whites as described above. Sprinkle with paprika. Garnish with small pieces of pimento, if desired.

Chimichurri Deviled Eggs

Take deviled eggs to new culinary heights by adding a verdant green chimichurri sauce to the mix. The sauce hails from Argentina and can easily be whipped up with ingredients most kitchens will have on hand. To make this sauce even more "green," consider picking up the fresh herbs from a local farmer's market, or, better yet, grow them yourself in containers or your backyard. *Yield: 24 servings*

YOU WILL NEED:

12	hard-boiled eggs
2	cloves of garlic, finely minced
1/2	teaspoon (3 g) sea salt
1	bunch parsley, stems removed
1	bunch cilantro, stems removed
1/8	ounce (3.5 g) fresh oregano leaves
1/4	of a medium sweet onion, finely diced
2	tablespoons (30 mL) lemon juice
1	tablespoon (15 mL) sherry vinegar
	Several dashes of hot sauce
1/2	cup (120 mL) olive oil
3	tablespoons (45 mL) mayonnaise

TO PREPARE:

1. Remove shells from eggs; rinse under cold water. Cut eggs in half lengthwise. Carefully remove yolks and transfer to a medium bowl; set whites aside.

2. Combine finely minced garlic with sea salt, and mince some more to combine. Set aside; the garlic will mellow as it sits.

3. Add parsley, cilantro, oregano, salted garlic, onion, lemon juice, sherry vinegar, and hot sauce to a food processor or blender. Puree for about 20 seconds. Add olive oil. Blend for an additional 20 seconds or so. This is your chimichurri sauce.

4. Add 5 tablespoons (75 mL) chimichurri sauce and mayonnaise to the reserved egg yolks. Mash yolk mixture till creamy.

5. Either fill a pastry bag and pipe filling into egg-white halves, or simply portion filling out with a spoon into each half. Top each deviled egg half with a dollop of the leftover chimichurri; refrigerate the remainder for another use. Serve immediately, or refrigerate for up to three hours.

Zucchini, Basil & Goat Cheese Frittata

What omelets are to the French, frittatas are to the Italians. Initially cooked in a skillet atop a stove, the frittata is finished under the broiler. Perfect if you are cooking for a group, as a frittata will finish cooking all at once. Not quite as finicky as omelets, frittatas can be served in wedges and are equally delicious hot or cold. The recipe below takes advantage of the summer abundance of zucchini and basil. Pick up some of each at either a local farmers' market or from your own backyard bumper crop. Vary the recipe to make use of what is available during other seasons. Asparagus and pea frittata would be magical in spring, while a mushroom and chard version takes advantage of fresh fall flavors. *Serves 6.*

YOU WILL NEED:

- 1 tablespoon (15 mL) extra-virgin olive oil
- 2 garlic cloves, minced
- 1/4 cup (56 g) red onion, minced
- 2 medium zucchini, sliced into thin rounds
- 6 large eggs
- 1/2 cup (120 mL) whole milk or heavy cream
- 1 tablespoon (15 g) butter
- 1/4 cup (56 g) fresh basil, chopped or torn into small pieces
- 1/2 cup (112 g) crumbled goat or feta cheese
- Sea salt
- Freshly ground black pepper

 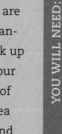

TO PREPARE:

→ 1. In a 10-inch (25.4 cm) ovenproof pan or skillet (preferably nonstick) warm olive oil over medium heat.

→ 2. Add onion, and cook until limp and translucent, about 4 to 5 minutes. Add garlic; sauté 3 minutes until brown. Be certain to stir every few minutes with a spatula to keep onion from sticking to the bottom of the pan. Add zucchini to onion mixture, and sauté 4 minutes. Transfer cooked vegetables to a small bowl. Set aside.

→ 3. Position the top oven rack about 6 inches (15.2 cm) beneath the broiler and preheat.

→ 4. In a medium mixing bowl, whisk eggs, a pinch of salt, and black pepper briskly until foamy. Add milk or cream, and whisk until incorporated.

→ 5. Add the butter to the same pan used to cook vegetables, turn heat to medium, and stir until melted. Pour the egg mixture into the pan.

→ 6. Turn heat down to medium-low and continue cooking until eggs are set on the bottom, about 4 minutes. The top will still be wet at this point. Sprinkle basil over eggs. Next add the garlic, zucchini, and onion mixture. Crumble cheese evenly over the top of the mixture.

→ 7. Place the skillet under the broiler. Broil, checking often, until the top of the frittata is golden and sizzling, 2 to 3 minutes.

→ 8. Remove from the oven. Either serve immediately in the skillet, or loosen with a spatula and slide the whole frittata onto a platter to serve.

Fresh Basil

Zucchini

Sweet Potato Soufflé

Tall, billowy, and light as a feather, few dishes pack as much dramatic punch as a soufflé. It's as if, once fused together, all the ingredients decided to collectively hold their breath, only to release it in one big "surprise!" moment. Many home cooks shy away from soufflé-making, perhaps on account of its theatrical nature. The truth is, beneath the sensationalistic façade lies a rather humble dish. What is essential, however, to pulling off a gorgeous soufflé, is to have your table set and diners assembled, forks at the ready, the moment you pull the dish from the oven. The grand entrance is the most enthralling part of the act; come late to the table and you just may miss it. The following recipe would be equally welcome at a holiday feast or a quiet midweek winter meal. While sweet potatoes are available year round, they are at their peak season during cooler months. Look for them at your local farmers' market. *Serves 4 as a side dish.*

YOU WILL NEED:

2	sweet potatoes, peeled and cubed
¼	cup (56 g) butter
¼	cup (60 mL) half and half
2	teaspoons (10 g) brown sugar
2	tablespoons (30 g) all-purpose flour
½	cup (120 mL) whole milk
3	eggs, separated, room temperature
⅓	cup (80 mL) maple syrup
¾	teaspoon (3 g) freshly ground nutmeg
	Pinch of sea salt

TO PREPARE:

1. Steam potatoes in a steamer basket over medium-high heat; when soft, transfer to a medium mixing bowl. Add half the butter and the half and half. Beat with a whisk until smooth. Set aside.

2. Adjust the placement of racks in the oven so that when the soufflé is centered on the middle rack, it will have adequate room to rise. Preheat the oven to 375°F (190°C). Using 1 tablespoon (15 g) butter, grease a 1½ quart (1.5 L) soufflé dish, gratin pan, or other baking dish with high, straight sides. Sprinkle brown sugar to coat the interior. Set aside.

3. Over medium heat, melt the remaining tablespoon (15 g) butter in a medium stainless-steel saucepan. Add flour and stir with a wooden spoon until golden, two to three minutes (this is your roux). Gradually add milk, whisking thoroughly between additions to prevent any lumps. Continue to whisk until thickened, about 2 minutes. Remove from heat.

4. Add the white sauce to the sweet potato mixture along with the egg yolks, maple syrup, and nutmeg. Stir until all ingredients are well incorporated.

5. In the bowl of an electric mixer fitted with whisk attachment, beat egg whites with sea salt until stiff peaks form. Be careful not to over-whip the whites. You want them to stiffen up, but still be moist and smooth in appearance. It may be necessary to stop and start your mixer during this step to get the whites to the right consistency.

6. Remove about a quarter of the whites and whisk them into the sweet potato mixture. Using a rubber spatula, carefully fold in the remaining whites, taking caution not to stir or beat, as doing so would cause the egg whites to deflate.

7. Transfer the mixture to the prepared dish. Place on a baking sheet and put in the center of the oven. Bake until puffy and golden, 40 to 45 minutes. Be certain not to open the oven door for any reason during the baking time. Serve immediately.

Berry Delicious Curd

No matter where you live, at some point in the year berries will be available, either for picking yourself or for purchase from your farmers' market or grocer. Petite, succulent, and packed with a tart sweetness, berries epitomize the saying "good things come in small packages." In this recipe, any combination of raspberries, blackberries, blueberries, boysenberries, or mulberries may be used, depending on what is available to you. Garnishing a bowl of ice cream, layered with pound cake, or served alongside freshly whipped cream, berry curd combines the perfect marriage of sweet and sour in every delicious mouthful! *Yield: 2 cups (480 mL)*

YOU WILL NEED:

- 3 cups mixed berries (672 g), fresh or frozen (thawed, if frozen)
- 1/3 cup (75 g) sugar
- 1/4 cup (56 g) butter, unsalted
- 1 1/2 tablespoons (23 mL) fresh lemon juice
- 2 eggs
- 2 egg yolks

TO PREPARE:

→ 1. In a food processor or blender, purée the berries. Using a mesh strainer placed over a medium bowl, strain berry purée, pushing on the mixture with the back of a spoon; discard seeds. Reserve 1 1/2 cup (360 mL) of the purée; save any remaining portion in the refrigerator for another use.

→ 2. In a medium, stainless steel pan, heat purée, sugar, and butter over medium-high heat. Stir in the lemon juice. In a separate bowl, gently whisk eggs and yolks. Whisk in about 1/2 cup (120 mL) of the warm purée into the egg mixture, then return the entire mixture to the pot of purée on the stove.

→ 3. Reduce heat to low and cook until slightly thickened, stirring constantly.

→ 4. Remove from heat, transfer to a glass or ceramic bowl, and chill covered in the refrigerator for at least six hours. Curd will thicken as it cools.

Pumpkin Crème Brûlée

What better way to welcome autumn and use up an abundance of eggs than with a warm, spicy, comforting pumpkin crème brûlée? Cupped in your hands at the completion of an autumn harvest meal or enjoyed outdoors around a campfire, this treat showcases the best of the season in every succulent bite. You can use either freshly steamed pumpkin or canned pumpkin for this recipe. Alternatively, butternut squash will work just as well. *Serves 6.*

YOU WILL NEED:

- 8 egg yolks
- ¼ cup (60 mL) maple syrup
- 1 cup (240 mL) heavy cream
- 1 cup (240 mL) whole milk
- ⅔ cup (160 mL) pumpkin puree
- ¼ teaspoon (1.25 g) nutmeg
- ½ teaspoon (2.5 g) cinnamon
- pinch of cloves
- pinch of allspice
- ¼ cup (56 g) light brown sugar, for topping

TO PREPARE:

→ 1. Put six ramekins or ceramic cups into a roasting pan or deep-sided pan. Set aside. Preheat the oven to 350°F.

→ 2. In a medium mixing bowl, whisk together the egg yolks and maple syrup.

→ 3. In a medium saucepan, gently warm the milk and heavy cream over medium heat until almost boiling. Remove from heat. Gradually whisk the hot milk into the eggs. Using a mesh sieve, strain the entire mixture back into the saucepan. Warm the custard gently over low heat, and stir in the pumpkin and spices.

→ 4. Pour the custard into the ramekins. Fill the roasting pan with warm water until halfway up sides of containers. Bake for 30 minutes until the custard centers are slightly soft. Allow to cool in the water bath, then remove the ramekins and refrigerate for at least four hours.

→ 5. Before serving, sprinkle brown sugar evenly over the tops. Caramelize with a blowtorch, or place in the toaster oven or under the broiler until the tops begin to turn brown and sizzle.

Resources

A NOTE FROM THE AUTHOR: It takes a good bit of time for a book to make its way from initial idea to printed copy. That means that as helpful as a list like this one is, by the time you're holding your book in your hands, it's inevitable that some of the website or physical addresses will have changed and that useful new resources will have surfaced. Still, it's a great starting point for those just entering the world of all things poultry. And about those changes and updates? I'm tracking them all in a regularly updated Resources section of my blog. Be sure to visit me regularly there: small-measure.blogspot.com.

POULTRY AND SUSTAINABLE LIVING MAGAZINES

Back Home
www.backhomemagazine.com

Backyard Poultry
www.backyardpoultrymag.com

Fancy Fowl (UK)
www.todaywebsitedesigns.com/ff-feb-welcome.html

Feather Fancier (Canada)
featherfancier.on.ca

Grit
www.grit.com

Hobby Farms
www.hobbyfarms.com

Living the Country Life
www.livingthecountrylife.com

Mother Earth News
www.motherearthnews.com

Poultry Press
www.poultrypress.com

Poultry Times
poultryandeggnews.com/poultrytimes/index.shtml

Practical Poultry (UK)
www.practicalpoultry.co.uk

POULTRY INFORMATION WEBSITES

Home Grown Poultry Magazine
homegrownpoultry.com

This site is all about marketing, showing, and raising different types of poultry.

BackyardChickens.com
www.backyardchickens.com

An extremely comprehensive site, with information for the novice and the experienced flock owner alike

ThePoultrySite.com
www.thepoultrysite.com

A website dedicated to information regarding poultry health

TheCityChicken.com
home.centurytel.net/thecitychicken/

A site for all things related to urban chicken enthusiasts

City Chickens
http://citychickens.com

A Seattle-based site for urban chicken raising

FeatherSite
www.feathersite.com

Comprehensive website with information on various types of poultry, with videos, photographs, and extensive zoological information

Poultry One
poultryone.com

This site contains articles, research, and a wealth of information for poultry enthusiasts.

Chicken Feed
lionsgrip.com/chickens.html

A highly informative site all about chicken feed, with information ranging from how to make your own feed to antiquated and traditional modes of feeding poultry. Also contains a state-by-state listing of organic feed producers.

HATCHERIES AND SUPPLIERS

Belt Hatchery
7272 S. West Ave.
Fresno, CA 93706
559-264-2090
www.belthatchery.com

Cackle Hatchery
P.O. Box 529
Lebanon, MO 65536
417-532-4581
www.cacklehatchery.com

Decorah Hatchery
406 W. Water St.
Decorah, IA 52101
563-382-4103
www.decorahhatchery.com

Double R Supply
5156 Minton Road Northwest
Palm Bay, FL 32907
dblrsupply.pinnaclecart.com

Egganic Industries
3900 Milton Hwy.
Ringgold, VA 24586
800-783-6344
www.henspa.com

Eggcartons.com
9 Main Street, Suite 1F
P.O. Box 302
Manchaug, MA 01526-0302
888-852-5340
Containers for selling and marketing eggs.

Eglu
86-OMELET-USA
www.omlet.us
Home of the Eglu housing unit.

Estes Hatchery
805 N. Meteor Ave.
P.O. Box 5776
Springfield, MO 65802
800-345-1420
www.esteshatchery.com

Heartland Hatchery
RR1, Box 177A
Amsterdam, MO 64723
660-267-3679
www.heartlandhatchery.com

Hoover's Hatchery
205 Chickasaw Street
Rudd, IA 50471-5025
800-247-7014
www.hoovershatchery.com

Meyer Hatchery
626 State Route 89
Polk, OH 44866
888-568-9755
www.meyerhatchery.com

My Pet Chicken
501 Westport Ave #311
Norwalk, CT 06851
888-460-1529
www.mypetchicken.com

Nasco Farm & Ranch
901 Janesville Avenue
P.O. Box 901
Fort Atkinson, WI 53538-0901
800-558-9595
www.enasco.com/farmandranch/

Privett Hatchery
P.O. Box 176
Portales, NM 88130
877-PRIVETT
www.privetthatchery.com

Sand Hill Preservation Center
1878 230th Street
Calamus, IA 52729
563-246-2299
www.sandhillpreservation.com
Purveyors of endangered and heirloom breeds

Welp Hatchery
P.O. Box 77
Bancroft, Iowa 50517
800-458-4473
www.welphatchery.com

Appendix

CHICKEN CARE CHECKLIST

While most care for your flock will occur on a day-to-day basis, some tasks are reserved for weekly, monthly, biannual, and annual attention. Daily and weekly tasks will quickly become habits, while those performed less often might need to be jotted down on your calendar. Although routine care doesn't necessarily guarantee your crew will never succumb to any unpleasantries, it sure goes a long way toward achieving that goal.

Daily

- Let chickens out of henhouse in the morning
- Fill waterers
- Remove frozen water during cold weather
- Fill feeders
- Remove any branches, leaves, or other matter from waterers and feeders as needed
- Remove eggs
- Store eggs, pointed end down
- Fluff up litter if using deep litter method
- Give a quick glance at your flock for any signs of distress or injury
- If there are any damp spots in the henhouse, remove them
- Lock chickens up at night

Weekly

- Scrub rim of waterer
- Clean feeder as needed
- Examine fence perimeter for any signs of burrowing
- If providing grit, refill supply as needed
- Add litter to nesting boxes or henhouse floor as needed
- Replace any soiled bedding, and empty droppings tray
- Scrape any droppings off of roosts

Monthly

- Check that roosts are sturdy and have no rough edges
- Examine entire coop for any signs of wear or rot, and repair as needed
- In the winter, check for drafts and remove source if found
- Check that all latches and locks are secure
- Place orders for feed and scratch or purchase from local feed store
- Purchase bedding material as needed
- Empty and refresh nesting box litter
- If needed, mow grass in run

Biannually

- If using deep litter method, empty out and compost all bedding; replace with fresh bedding
- Check for leaks in roof and repair as needed

Annually

- Remove all bedding, feeders, waterers, and nesting materials
- Disinfect living quarters with a 1:10 bleach and water solution administered in a spray bottle
- Scrub floor with disinfectant solution
- Allow everything to dry completely before putting materials back in
- Add fresh bedding material to floor and nesting boxes
- Put a fresh coat of paint on coop, if needed
- Repair any openings or worn parts of fencing

Glossary

Albumen. The transparent protein surrounding an egg yolk, otherwise known as an egg white

Air sac. An airspace found inside the rounded end of a fertilized egg; it helps in retaining moisture, assisting in embryo development.

Bantam. A small breed of chicken, whose size is roughly one-quarter that of a standard-size chicken; while most bantams are merely scaled-down versions of their larger counterparts, several breeds have no larger version and are referred to as "true" bantam breeds.

Bedding. The absorbent material used to line coop floors and nesting boxes, also referred to as "litter"

Bloom. The protective coating on a freshly laid egg that seals its pores, thereby keeping bacteria from getting in and moisture from escaping

Breed. A group of chickens with related physical attributes, such as comb, shape, and plumage; also used as a verb to describe the act of mating between a hen and a rooster to create fertilized eggs

Broiler. A young bird intended for table to be served in parts; also referred to as a "fryer"

Brood. Used as a verb to describe the act of a hen sitting on eggs, attempting to hatch them, a brood is also used as a noun to describe the hatched chicks themselves. A hen either sitting or displaying characteristics of a desire to sit is referred to as a "broody" hen. The heated enclosure used to keep chicks warm in lieu of a hen is a "brooder."

Cannibalism. In relation to chickens, the acts of pecking one another's feathers, body parts, or eggs

Candle. A technique used to examine the contents of an egg, performed by shining bright light through it; the device used in producing light is called a "candler."

Cloaca. A chamber inside a chicken where the urinary, digestive, and reproductive systems meet and open to the vent

Clutch. A group of eggs that are hatched together

Cock. A male chicken at least one year of age or older; more commonly referred to as a "rooster"

Cockerel. A male chicken less than one year old

Comb. The red, fleshy protrusion topping a chicken's head

Coop. The enclosed area in which a flock of chickens live; both indoor and outdoor quarters are often collectively referred to as the "run."

Crest. The outcropping of feathers adorning the heads of some breeds of chickens, such as Polish, Crèvecœur, Houdan, and Silkie; also called a "topknot"

Crop. A pouch inside a chicken's esophagus where food is stored and softened before moving on to the rest of the digestive system; often visible as a bulge in a chicken's chest at the end of the day

Crossbreed. A chicken whose parents are of two different breeds

Cull. The act of removing a chicken from your flock, either through butchering or relocation to another home

Droppings. Chicken poop

Dual-purpose breed. A type of chicken kept for both eggs and meat

Dust bath. An area of dry, sandy soil in which a chicken will root around, spread dirt over herself, and rest; assists in killing mites and lice

Embryo. An unhatched fertile egg

Flock. A group of chickens who inhabit the same living quarters

Free-range. Chickens permitted to forage and pasture without confinement

Gizzard. The part of a chicken's digestive system where grit is stored and food is ground down

Grit. Small pebbles, sand, or other stony material eaten by chickens and used to grind up food in the gizzard

Hatch. The act of live chicks emerging from their shells

Hen. A female chicken at least one year old

Hybrid. Chickens bred from two different breeds to produce desired characteristics, such as egg-laying ability or meat production

Incubate. The act of creating favorable atmospheric conditions for successfully hatching eggs

Layer. A chicken kept for purposes of egg production

Litter. The absorbent material used to line coop floors and nesting boxes, also referred to as "bedding"

Meal. Coarsely ground grains given as food to chickens, often with added supplements; also referred to as "ration"

Molt. The annual shedding and renewing of feathers; done twice during a chicken's first year of life

Nesting box. The location in which a chicken lays eggs; ideally, it should be situated in the darkest, quietest part of the henhouse.

Oviduct. A tube-shaped reproductive organ inside a hen through which eggs pass en route to the vent

Pasting/pasty butt. A life-threatening condition most often affecting chicks in which the vent becomes crusted over with droppings

Pecking order. The hierarchy naturally established within a flock of chickens

Perch. The ledge or pole on which a chicken sleeps, also called a "roost"; may also refer to the act of standing on a roost

Plumage. The feathers covering a bird's body

Point of lay. The time at which a young female could begin laying eggs, usually around 18 weeks of age

Pullet. A female chicken less than one year old

Purebred. A breed that has not been crossed with any other breed; a chicken whose parents are the same breed

Ration. Coarsely ground grains given as food to chickens, often with added supplements; also referred to as "meal"

Roaster. Chickens raised for table to be served whole

Roost. The ledge or pole on which a chicken sleeps, also called a "perch"; may also refer to the act of standing on a roost

Rooster. A male chicken at least one year of age or older; also referred to as a "cock"

Scratch. Whole cereal grains fed to chickens

Sexed chicks. Newly hatched chickens whose sex is determined prior to purchase

Shank. The lower part of a chicken's leg, located between the thigh and the foot

Spur. The sharp, pointy protrusions found on the backside of a rooster's shank

Standard. A chicken meeting the ideal characteristics for its breed; also used to refer to those characteristics as described in the Standard of Perfection

Straight run. A group of newly hatched chicks whose sex is undetermined at purchase

Vent. The opening at the rear of a chicken, through which eggs and excrement are eliminated

Wattles. The two red flaps of flesh dangling from either side of a chicken's chin

Acknowledgments

An enthusiastic shower of gratitude is in order for the many talented individuals who pulled together this project.

To the wonderful profilees: Kevin Bartoy, Patrick Barber and Holly McGuire, Robin Cape, Theresa Freeman, Zev Friedman, Erik Knutzen and Kelly Coyne, Christine Kaltoft, Natasha Shealy, Jenny Mercer, and Lynne and Bruce Van Dyke-Grammer. Thank you for putting a face to the many different ways we can all welcome chickens into our lives.

Abundant thanks are offered to the Graves Family of Double G Ranch for sharing with us their gracious hospitality. The pages of this book are richer because you opened your home and farm to us. Your children, Jessica and Dalton, demonstrate just how well chickens can become a valuable part of the family and teaching tool.

I applaud Rebecca Springer for waving her copyediting wand over my words with grace and good judgment.

I can't say enough about the phenomenal talent of the book's strong visual team. Photographer Lynne Harty never ceased to surprise with her ability to get just the right shot over and over again, even when it involved getting very up close and personal with more than a few of our feathered models. Designer Eric Stevens should be lauded for his beautiful design, illustrations, and flawless ability to pull together all of the puzzle pieces in just the right way. Melanie Powell's artful drawings will aid even the most DIY challenged in creating the nesting boxes and chicken tractor, while Orrin Lundgren's added illustrations put the final important touches of clarity to the nitty-gritty details.

Heartfelt thanks to Nicole McConville, for believing I had both the skills and gumption to take on this book and the series. I am immeasurably grateful to have such a great friend and editor wrapped up in one convenient package.

For my long-suffering, infinitely patient husband, Glenn, who served as chef, woodworking consultant, cheerleader, and therapist through all of this, I appreciate you more than you could ever know.

Finally, special thanks to Meaghan Finnerty, Chris Bryant, Paige Gilchrist, and Marcus Leaver for all of the excitement surrounding the series that you have both nurtured and enabled.

Also Available in the Homemade Living series:

AVAILABLE SPRING 2011:
Home Dairy with *Ashley English*
Keeping Bees with *Ashley English*

Photo Credits

The pages of this book are richer thanks to the contributed photos. Much gratitude is owed to the following individuals: Patrick Barber (pages 19, 31, 42, 50, 51), Jennifer Bartoy (page 99), Aleta Belden (page 107), Laura Carmer (page 91), Cathy Caspersz (page 43), Caroline Clerc (page 63), Candace R. Coefield (page 28), Robert Collins (page 29), Dean Cully (page 31), Ingrid Douglas (page 28), James Hood (page 35), Irene Kightley (page 27), Nicole McConville (pages 24, 73), Holly McGuire (page 19), Melissa Mills (page 13), Marnie Muller (page 41), Julie Olsen (page 26), Melinda Seyler (pages 42, 50), Olivia Shealy (page 87), Mary Simmons (page 30), Tim vonHolten (page 50), Lisa West (page 27), Thomas Brent Wilson (page 17), and Daniel Worrall (page 32).

Index

Index (continued)